Advances in Geographical and Environmental Sciences

Advances in Geographical and Environmental Sciences synthesizes series diagnostigation and prognostication of earth environment, incorporating challenging interactive areas within ecological envelope of geosphere, biosphere, hydrosphere, atmosphere and cryosphere. It deals with land use land cover change (LUCC), urbanization, energy flux, land-ocean fluxes, climate, food security, ecohydrology, biodiversity, natural hazards and disasters, human health and their mutual interaction and feedback mechanism in order to contribute towards sustainable future. The geosciences methods range from traditional field techniques and conventional data collection, use of remote sensing and geographical information system, computer aided technique to advance geostatistical and dynamic modeling.

The series integrate past, present and future of geospheric attributes incorporating biophysical and human dimensions in spatio-temporal perspectives. The geosciences, encompassing land-ocean-atmosphere interaction is considered as a vital component in the context of environmental issues, especially in observation and prediction of air and water pollution, global warming and urban heat islands. It is important to communicate the advances in geosciences to increase resilience of society through capacity building for mitigating the impact of natural hazards and disasters. Sustainability of human society depends strongly on the earth environment, and thus the development of geosciences is critical for a better understanding of our living environment, and its sustainable development.

Geoscience also has the responsibility to not confine itself to addressing current problems but it is also developing a framework to address future issues. In order to build a 'Future Earth Model' for understanding and predicting the functioning of the whole climatic system, collaboration of experts in the traditional earth disciplines as well as in ecology, information technology, instrumentation and complex system is essential, through initiatives from human geoscientists. Thus human geoscience is emerging as key policy science for contributing towards sustainability/survivality science together with future earth initiative.

Advances in Geographical and Environmental Sciences series publishes books that contain novel approaches in tackling issues of human geoscience in its broadest sense — books in the series should focus on true progress in a particular area or region. The series includes monographs and edited volumes without any limitations in the page numbers.

Utsav Kumar Singh

Regional Dimensions of Human Development in India and South Africa

Through Sustainable Development Goals

Springer

Utsav Kumar Singh
Institute of Social Sciences
New Delhi, Delhi, India

ISSN 2198-3542 ISSN 2198-3550 (electronic)
Advances in Geographical and Environmental Sciences
ISBN 978-981-97-5845-6 ISBN 978-981-97-5846-3 (eBook)
https://doi.org/10.1007/978-981-97-5846-3

This Springer imprint is published by the registered company Springer Nature Singapore Pte Ltd.
The registered company address is: 152 Beach Road, #21-01/04 Gateway East, Singapore 189721, Singapore

To Rachana

Foreword

There is little dispute that economic growth and human development are deeply intertwined. And yet the nature of relationship between these two critical outcomes is hotly disputed by scholars, journalists, and activists alike. Nowhere is this truer than in India, where prominent neo-liberal scholars such as Arvind Panagariya and well-known thinkers associated with left, notably Amartya Sen, have clashed over the role of the state in development. Should India, and emerging economies more generally, focus on promoting growth through the cultivation of free markets and pro-business policies? Or should the country rather emphasize a form of growth 'with adjectives', one that is sustainable, widely inclusive, human-centred, even redistributive? This debate is not just academic; it touches the heart of growth strategy debates around the world and, indeed, has implications for the lives of billions of people.

Utsav Kumar Singh's book, with its particular combination of academic rigour and moral passion, is not afraid to take on this debate, and indeed many others. Singh's starting point is a comparative analysis of India and South Africa, by applying the United Nations Sustainable Development Goals (SDGs), something rarely attempted by other scholars. He leverages this comparison to explore why India, with its greatly superior growth rate, is nevertheless falling well behind South Africa in human development indicators. It might be expected that India, with nearly 80 years of independence and a long history of social democracy, would prioritize, for example, education and health for the poor. It might furthermore not be a surprise if South Africa, emerging from decades of oppressive white rule, would score poorly in delivering critical services to its marginalized citizens. And yet ...

I first met Dr. Singh at a conference on Indian federalism held in New Delhi in 2022. What immediately struck me was his passion for improving the lives of his fellow citizens and, especially, his fellow Biharis. Utsav is a proud son of Bihar, and his love for the state, combined with his disappointment in the pace of its progress, motivates much of his work. Utsav's previous book explores the case of Bihar directly, but this one takes a broader look at how countries can promote human development even in chastened circumstances.

In the final analysis, Dr. Singh shows that India, and other governments facing similar material circumstances, can reach the SDGs without sacrificing traditional economic growth. Utsav reminds us that human development, whatever its cost in the short term, is critical for long-run growth. After all, a country cannot expand its economy for long without an educated and healthy workforce. This is good news for economic policymakers, at least for those willing to take the long view. Yes, economic growth is a must for emerging economies, but it can be pursued, indeed must be pursued, in the context of policies that promote human flourishing.

March 2024

Charles Hankla
Director, International Center for Public Policy, Public Finance Research Cluster, Andrew Young School of Policy Studies; Professor of Political Science
Georgia State University
Atlanta, GA, USA
chankla@gsu.edu

Preface

> *Despite unprecedented increase in overall opulence, the contemporary world denies the elementary freedom to vast numbers—perhaps even the majority—of people...*
>
> *—Amartya Sen.*

In *Development as Freedom* Amartya Sen argues that freedom is central to the ends and means of development. He expanded the concept of freedom to include the capability to realize or prioritize functioning they have reason to value.

We are living in an era of unprecedented global integration. Unarguably, for many reasons, our life is better than our ancestors. People are enjoying a healthy and longer life than ever before. Liberal constitutions protect our rights and people have increasing freedom and avenues for the pursuit of happiness. However, human advancement has limitations and is prone to negative externalities. Recently we have experienced one such, the COVID-19 pandemic. We witnessed the unparalleled global growth halted by the swift spread of the deadly virus. The pandemic also unmasked the existing inequalities in societies. The unprecedented challenges of COVID-19 and its after-effects have pushed the world into a situation of uncertainty.

In the post-COVID era, India emerged as the fastest growing economy in the world. Buzz about India's stupendous economic growth rate and its vast human power especially the potential of a huge youth population has become public discourse. But there is a snag. India is home to the world's largest poor population and almost sixty percent of its population depends on subsidized food provided by the government. This enigma draws attention to Sen's idea of development.

I began my study by examining the Indian economy since independence. In the age of neo-liberalism, the world experienced two major negative externalities: the North Atlantic Economic Crisis and the COVID-19 pandemic. Therefore, testing of hypothesis in a global preview necessitated the selection of another region with similar conditions. After evaluating several countries, especially developing countries, I choose South Africa. India and South Africa share many commonalities, from the colonial past to the Gandhian approach of the liberation movement. South Africa is also an enigma. It is the world's most unequal country with half of the population living in poverty while scoring a good rank in development parameters.

Although India has achieved impressive economic growth in recent decades, the country performed weakly in the manufacturing sector. Despite the improvement in poverty reduction, India is still home to the largest poor population in the world. Similarly, the highly industrialized economy of South Africa also faces the challenge of high unemployment, poor quality of primary education, low-skilled labour force, poverty, and persistent inequality.

The persistent poverty levels indicate that drivers of upliftment such as access to good health, quality education, skill, clean and sustainable water, and sanitation services are not reaching everyone. On the contrary, despite the sluggish economic growth rate, South Africa performed better than India in major global development indicators. For instance, in the Human Development Index Report-2022, India's position was 132 while South Africa stood at 109. India's performance in the Global Hunger Index-2023 was pathetic. In the list of 125 countries, India stood at 111th while South Africa was ranked 58th. The Multi-dimensional Poverty Index-2023 shows that 16.3 percent of India live under multidimensional poverty whereas the corresponding figure for South Africa is 6.3 percent.

While comparing India and South Africa, I turn to Sen's idea of freedom and its role in enhancing human capabilities. His idea was mirrored in the vision of Pt. Jawaharlal Nehru and Nelson Mandela, the founding fathers of India and South Africa, respectively. For both leaders, freedom entailed a step forward in eradicating poverty and ignorance, reducing the burden of diseases, and ensuring equal and unbiased opportunities for all. The vision of both also resonated with the philosophy of human development.

Human development is the most important question faced by modern society. Poverty and global inequality have reached a disturbing level. Development must be able to free the extremely poor from their hopeless miseries. From the lens of human development, poverty is the limitation on the individual in making choices and denial of opportunities for a reasonable life. The multiple dimensions of poverty are encapsulated in Sustainable Development Goals (SDGs), a set of 17 goals with 169 targets to be achieved by 2030. SDGs are in turn an expansion of the Millennium Development Goals (MDGs) to finish the incomplete task through comprehensive and relevant indicators and provide a way forward to a humane world. SDGs are interlinked and balance the social, economic, and environmental dimensions of development. The attainment of SDGs in turn depends on policy design for human development.

The progress towards SDGs has been immensely affected by the COVID-19 pandemic. The impact of the pandemic was particularly harsh on the poor. The COVID-19 crisis has derailed the progress achieved in providing basic necessities and the developing world has realized the immediate need to invest in health and other infrastructure. Findings suggest that with improvements in health, education, and income, a rise in purchasing power can be anticipated on the demand side. Public spending on education and health is key to realizing human development. The government should maintain a balance between the states/provinces in the area of human development. Any government must provide health, education, food, shelter, and decent jobs to live a better life. Hence, the programme and policies of India and South Africa must further focus on the need to strengthen the determinants of human development to attain the Sustainable Development Goals by 2030.

I took this issue in my post-doctoral research at the University of Delhi under the supervision of Prof. Rekha Saxena. Her expertise in multilevel governance was instrumental in conceptualizing the idea that 'one size is not fit for all'. My post-doctoral research provided thc foundational base of the present book. Mainstreaming of gender, indigenous people, and other marginalized groups continues to define the way Indian and South African societies are imagined. The persistence of poverty coupled with hunger in India and South Africa has doleful consequences on decent income. This book traverses the stages in public policy from planning to implementation, and from reach to range, in the quest of attaining human development.

Despite the high growth rate, why is India's performance lower than South Africa's in terms of social indicators? This book addresses this puzzle by drawing attention to welfare approaches. This book offers a comparative study of the progress made by two regional economies with many similarities—India and South Africa—in realizing human development within the framework of Sustainable Development Goals (SDGs).

This book argues that uneven development challenges need asymmetric strategies, contrary to the axiom that SDGs are indivisible. India and South Africa are countries with unique federal structures that uphold the unbalanced states together. The further advancement of India and South Africa depends on transforming their quantity (large youth population) into quality (human capital). Both countries are also striving to create platforms to unleash human potential, which is critically examined. Both governments face the unique challenge of maintaining a balance between the uneven states/provinces while ensuring health, education, food, shelter, and decent jobs through targeted pro-poor policies.

This study is an attempt to empirically investigate the relationship between selected UN SDGs and GDP growth rate as its proxy in India and South Africa. It attempts to empirically investigate through multiple regression analysis. The central focus throughout the book is the investigation of the progress made by India and South Africa in Human Development. It examines the progress made by India and South Africa in achieving MDGs and then examines the gaps in the transition from MDGs to SDGs. It also critically discusses the strategies adopted by the Governments of India and South Africa to address the challenges.

This book attempts to establish the synergy between social, economic, and environmental goals in the Sustainable Development Goals (SDGs). It postulates that a better economic indicator is routed through strong social and environmental indicators. SDGs represent the next stage in the journey of human development among nations. Therefore, the countries need to move beyond MDGs towards SDGs so that the benefits of socio-economic development reach future generations. India and South Africa, as a signatory to the adoption of UN's 17 Sustainable Development Goals, are under obligation to streamline development policies for a time-bound (by 2030) achievement of human well-being. This book examines the progress and challenges in the way of two emerging economies towards SDGs.

New Delhi, India Utsav Kumar Singh

Acknowledgment

I was very fortunate to have a wonderful supervisor during my post-doctoral research, on which this book draws. Professor Rekha Saxena has not only supervised the research, but she has always been a source of motivation, inspiration, and encouragement. This work would not have been possible without her input, guidance, and consistent support which I was always able to count on throughout my association with her. I am grateful to her not only for her academic support but also for her genuine concern regarding my well-being in general. I am equally grateful to the Indian Council of Social Science Research (ICSSR) for providing a Post-Doctoral Fellowship, which helped me in bringing this work to completion.

I am grateful to Prof. Charles Hankla, Georgia State University, Atlanta, GA, for his graciousness in reading the manuscript and writing a foreword to this book, despite his busy schedule.

I thank Thejalhoukho Casavi for taking the burden to read the whole manuscript and correcting the proof. His assistance was crucial in formalizing this book. I have also benefitted from the insights of Abhishek Meena regarding the needs and demands of marginalized sections of Indian society. I am grateful to Jean Dreze for discussing the challenges for human development in India over a cup of tea at Gwyer Hall Canteen, University of Delhi.

Late Prof. R. B. Singh was instrumental in guiding me with the apt methodologies in pursuit of my questions. I am thankful to Prof. Subhash Anand and Prof. Pankaj Kumar of Geography Department, University of Delhi for reading my book proposal. I would like to thank Prof. Evgenia Kuznetsova, Ural Federal University for providing me a platform to discuss the post-pandemic issues and challenges in the developing world in the scintific seminars in the field of Social Policy at School of Public Administration and Entrepreneurship, Ural Federal University, Yekaterinburg, Russia.

My research is interdisciplinary in approach and I am indebted to all the help and assistance that I received from experts in various disciplines and insights of stakeholders. I have been very blessed to receive support from Gyaneshwar Singh (DDG, NCB, Ministry of Home, Government of India), Saurav Kumar Shah (Executive Director, Power Finance Cooperation Ltd.), Gyanendra Kumar Singh (ZD, NCB,

Delhi Zone), Ashok Basoya (AAG, Chhattisgarh), Dr. Ragini Nayak, and Ranjan Abhishek (Ministry of Education, Government of India).

I am grateful to Dr. Parmeet Kumar Vinit, Assistant Professor, Patna University for his assistance in regression analysis. My gratitude is also due to Ravi Tomar (DIT, Dehradun) for designing the illustrations in this book.

Sincere gratitude to Alisha Dhingra, (Satyawati College, University of Delhi), Mritunjay Kumar (NLU, Shimla), Rauhil Srivastava (CIPOD, JNU), Suryansh Kharb (University of Delhi), and Deepak Pathak (IDFC First Bank) for their invaluable contributions in formalizing this research.

This research was, to a great extent, shaped by the active class discussion with my students Swastik Singh, Subhra Ananad, Revanta Khanna, Ananya Arora, Amulya Gupta, Ananya Singla, Mohul, Surbhi Gupta, Raunit Gautam, Aaswash Mahanta, Akshit, Rishika, Kenneth, Shirley, Harsh, Satyam and others at Shaheed Bhagat Singh College, University of Delhi. I am thankful to my students for giving fresh perspectives on contemporary trends, issues, and challenges and I also thank my fellow researchers for engaging in constructive debates on various development issues over tea in the canteen.

I would like to thank my family members who were always pillars of perseverance and endurance all through my life and have never let me down in any moment of crisis. I dedicate this thesis to my beloved parents whose constant love, affection, and blessings encouraged me to carry out this work.

Finally, my inexpressible gratitude to my better half Rachana Tomar who is always my unfailing source of strength and comfort in all my pursuits despite her own arduous research engagement. Her unshaken belief in my research work has been biggest motivation. Words would not do justice so I leave it that way.

Utsav Kumar Singh

About This Book

This book offers a comparative study of the progress made by two regional economies with many similarities—India and South Africa—in pursuit of human development by applying Sustainable Development Goals (SDGs). In this effort, book traverses the stages in public policy from planning to implementation, and from reach to range, in pursuit of inclusive growth and well-being for all. The book includes cross-national qualitative as well as quantitative studies and deeper analyses of India and South Africa in respect of long-term development goals.

This book attempted to solve the following puzzle: despite the stupendous growth rate, why is India's performance lower than South Africa's in terms of human development indicators? This book addresses this puzzle by drawing attention to welfare approaches. These two countries have unique federal structures that uphold the unbalanced states together, and this book argues that uneven development challenges need asymmetric strategies, contrary to the axiom that the SDGs are indivisible. Further advancement of India and South Africa depends on transforming their quantity (large youth population) into quality (human capital). Both countries are also striving to create platforms to unleash human potential, which is critically examined.

Contents

Abbreviations

AfDB	African Development Bank
ANER	Adjusted Net Enrolment Ratio
ARHC	Affordable Rental Housing Complexes
BPL	Below Poverty Line
DGGHE	Domestic General Government Health Expenditure
DISCOM	Power Distribution Company
DPME	Department of Planning, Monitoring and Evaluation
EPWP	Expanded Public Works Program
FAO	Food and Agriculture Organizations
GDP	Gross Domestic Product
GEAR	Growth Employment and Redistribution'
GER	Gross Enrolment Ratio
GGHE	General Government Health Expenditure
GNP	Gross National Product
GPI	Gender Parity Index
HDI	Human Development Index
HDR	Human Development Report
IBC	Insolvency and Bankruptcy Code
IFAD	International Fund for Agricultural Development
IFC	International Finance Corporation
ILO	International Labour Organizations
IMF	International Monetary Fund
IPC	Indian Penal Code
IPU	International Parliamentary Union
ISRDS	Integrated Sustainable Rural Development Strategies
KPSS	Kwiatkowski-Phillips-Schmidt-Shin
MDGs	Millennium Development Goals
MGNREGA	Mahatma Gandhi National Rural Employment Guarantee Act
MIS	Multiple Indicator Survey
MMR	Maternal Mortality Rate
MOSPI	Ministry of Statistics and Implementation

MPI	Multidimensional Poverty Index
MSME	Micro Small and Medium Enterprises
MTSF	Medium Term Strategic Framework
MUDRA	Micro Unit Development & Refinancing Agency
NAS	Achievement Survey
NBFC	Non-Banking Financial Companies
NDP	National Development Plan
NEET	Not in Education, Employment and Training
NEP	National Education Policy
NFHS	National Family Health Survey
NIRF	National Indicators Framework
NISED	National Integrated Small Enterprise Development
NMR	Neo-Natal Mortality Rate
NPC	National Planning Commission
NSS	National Sample Survey
ODF	Open Defecation Free
PCGS	Partial Credit Guarantee Scheme
PCPNDT	Pre-Natal Diagnostic Techniques Act
PLFS	Periodic Labor Force Participation Survey
PMKVY	Pradhan Mantri Kaushal Vikash Yojana
PMUY	Pradhan Mantri Ujwala Yojana
PPP	Purchasing Power Parity
PSET	Post-School Education and Training
RDP	Reconstruction and Development Program
RGGVY	Rajiv Gandhi Grameen Vidyutikaran Yojana
RISD	Regional Indicative Strategic Development
RTE	Right to Education
SACU	Southern African Custom Union
SADC	Southern African Development Community
SC	Schedule Caste
SDGs	Sustainable Development Goals
ST	Schedule Tribe
STATS SA	Department of Statistics, South Africa
U5MR	Under 5 Mortality Rate
UHC	Universal Health Coverage
UN	United Nations
UNCED	U.N. Conference on Environment and Development
UNDP	United Nations Development Programs
UNFPA	United Nations Populations Funds
UNICEF	United Nations International Children's Emergency Fund
USD	United States Dollar
VGF	Viability Gap Funding
WFP	World Food Program
WHO	World Health Organizations

List of Figures

List of Illustrations

List of Tables

Chapter 1
Human Development in Paradigm of Sustainable Development Goals

Introduction

Some 2500 year back, the quest for human development can be seen in the Buddha's encounter with disease, poverty, hunger and other forms of human deprivation which forced him to think about the truth of life. Through his engagements with the deepest questions of existence he laid down a path of living which continues to impact lives across the globe. Buddha is unarguably the greatest hero that ever came from Indian soil.

In the journey of humanity, many Indian and western philosophers have paid significant attention to inequality dating back to as early as Kautilya and Ashoka in ancient India, and Adam Smith, Condorcet, Mary Wollstonecraft, Karl Marx or John Stuart Mill in the modern west. Rangarajan (1987) includes that in 4th Century BC Kautilya advocated the welfare of animals as well as welfare of humans. Scholarship of Long (2021) suggest that Ashoka was the champion of universalism, his rock edicts reflect his idea of peace, social welfare, environment conservation, political pluralism and compassion for justice. Wollstonecraft (1792) wrote worlds wants justice, not charity. In same year, Paine (1792) pointed a question, if the Christ genealogy traced to Adam, why the rights of man not trace to the creation of man? Further, he argued that all men are born with equal natural right, every child born today enjoy the same natural rights, available for the first child who born. Same idea resonates in Marxian philosophy who finds justice in equality which is further taken by mills in idea of liberty. Adam Smith recognized the dignity of human being and the importance of his ability to participate in social activities. Smith's notion of human dignity was also reflected in John Rawls account of 'primary goods' along with opportunities, rights and liberties, income and wealth. Amartya Sen expanded the concept of freedom to include the capability to realize or prioritize functioning they have reason to value. Anand and Sen (1994) found a common feature in mentioned ideas, 'Universalism', the core of policies meant for welfarism. It is seminal to the concept of 'human

U. K. Singh, *Regional Dimensions of Human Development in India and South Africa*, Advances in Geographical and Environmental Sciences,
https://doi.org/10.1007/978-981-97-5846-3_1

development', in a finite world, which is struggling to erase exiting inequities and conserving the nature for future generation.

The ideas of Marry Wollstonecraft, Thomas Pine, Karl Marx, Stuart Mills, John Rawls and Adam Smith are applied on each and every human being without any biasness based on caste, creed, religion and region. The idea upholds the core of human development. Ongoing debates in public policies must accommodate the feature of universalism coupled with sustainability. The approach defined as sustainable development, a belief that our future generation needs same attention that our present generation is receiving.

Approaches to Development

Development is the process that allows people to realize their human potential facilitated by socio-economic changes. Generally, the concept of development is essential to economics. In the first instance, economic problem involves 'planning' issues, and more of it unarguably on 'adjustment' of one kind to another. Secondly, the achievement of all this has been ultimately measured in terms of its contribution to enhancing human life. Enhancement of human life must be vital to the entire economic process and thus it is integral to the concept of development.

Growth Approach

Theories of development economies emerged as a subject of their own after the end of the second world war. Earlier scholarship suggests that in the aftermath of the war, policymakers across the world were trying to achieve rapid economic growth by expanding the GNP and full employment (Rosenstein-Rodan, 1943). Mandelbaum (1945) figured out the existence of discrepancy in per capita income in agriculture and industries at the same time since most of the population is engaged in agriculture, especially in backward countries. In the case of India, the country is in a stage of economic development but not in terms of per capita income development (Dasgupta, 1954). Economic development has much to do with human upliftment. Nurkse (1961) argued that capital is vital for economic development but not seminal to progress. Capital formation must include human capital created by investing in humans through health, education, and skills. With existing commonalities in the process of betterment of human life, backward and prosperous countries might be differed in planning and adjustment (Bauer, 1971). Nordhaus and Tobin (1973) argued that the goal of economic activity is to expand consumption rather than limiting to production, but 'GNP is an index of production, not consumption.' At the same time, the factors of production presage significant pressure on the biosphere, industries draw their raw material from natural sources (sea, soil, forest, and waterways). Although economic growth has successfully reduced infant mortality and enlarged the human life span, increased literacy rate, and child education, while food production is surpassing the population growth, the cost of these gains is environmental loss. Economic growth is concentrated in developed countries, it failed in containing the gap between developed and backward countries. The absolute number of incidences of hunger is rising in recent years, and the proportion of homeless and dependency on fossil fuel for cooking and other household chore is increasing.

The growth approach has its own shortcoming. In 2008, owing to the dissatisfaction of GDP statistics information, French President Nicholas Sarkozy established a 'Commission on the Measurement of Economic Performance and Social Progress' under the able leadership of some of the finest economists of our era, Prof Joseph. E. Stiglitz, Prof. Amartya Sen and Prof. Jean Paul Fitoussi. The Commission explored the limitation of GDP in measuring societal wellbeing. The findings suggests that policies should be targeted at expansion of human wellbeing, not GDP. For example, GDP measurement exclude social indicators such as level of poverty, inequality, unemployment among others, further, it does not address environmental loss (Stiglitz et al., 2009; Frieden, 2024). Joseph Stiglitz (2009) in his essay 'GDP Fetishism' challenged the GDP statistics, he argued that in many cases performance of an economy is far better than peoples' own perceptions, and questioned whether it is an appropriate measure for policy making. The GDP concept is useful to understand the fluctuation, but failed in explaining its contribution to the welfare of human beings.

The growth model assumes that the expenses of investments in development are incurred by the expansion of the economy. Consequently, market forces are becoming more determinant in influencing the trade-offs, for instance between environment and job creation. Thus, the definition of development is not universal.

The planning and adjustment for development process are dynamic, to some extent depending on circumstances, that might change from place to place and from people to people. But, the means of attaining development very much stands on three basic pillars; good health, better education which leads to a decent income. Further, development enriched by active participation in social and political activities and access to all types of opportunities available for the betterment of human life without any discrimination.

Thus, the world at that time was enthusiastic about a universal definition of development. A concept that is not restricted to monetary change, but also addresses the contribution of economic growth in raising the living standard of people and conserving the biosphere. In 1987, the world leaders univocally agreed to the concept of 'sustainable development', a universal definition of development that emerged in the communique of the Brundtland Conference.

Sustainable development is a new approach to sustaining economic growth without compromising the well-being of future generations by conserving natural resources. It is a call for a common future that is more lively, more secure, and more just. To realize this concept, it is important to identify the challenges that the world is confronting, and establish a framework to manage natural resources and sustain 'human development.' To make the idea of sustainability more rational, and less emotional, agencies could facilitate policy debate which leads to common agenda to protect the earth (Solow, 1992). Giannetti et al. (2015) suggested to switch the development measure from economic growth to human wellbeing and sustainable development.

Human Development Approach

The range of GDP does not cover many things that hold value as they can only be measured by parameters of liberty, health, education and the like. A distinction between 'commodities' and 'capabilities' had been given by Amartya Sen in the 1980s. The commodities show up in the GDP while the capabilities do not. Mahbub Ul Haq and Amartya Sen in a spearhead project developed new approach to measure the development of people in a country; the human centric approach based on Human Development.

Human development, as an approach, is related to the fundamental idea of development. It is all about expansion of human life, rather than expansion of the economy, which accommodate human life, which is only a medium to it. Human development is process of advancing the richness of people choice.

Human development is the most important question faced by the modern society. Poverty and global inequality have reached a disturbing level. Development must be able to free the extremely poor from their hopeless miseries. Development should respond concretely to people's needs. The 1990 Human Development Report considers people as the real wealth of a nation (HDR, 1990). So, people are not just the means but the principal end of development.

Development should be seen as the extension of the set of attainable and worthy life options (Gasper, 2002). This approach is influenced by the work of Amartya Sen. The human development paradigm of UNDP, with its emphasis on valuable life options, is based on Sen's Capability Approach.

Sen (1979, 1979c, 1979a, 1979b) states that welfare economics evaluates human well-being through indices such as the Gross National Product, the Gross Domestic Product, and per capita income; at the same time, these concepts miss out on the other crucial aspects of human well-being such as human health, human education, and human capabilities to make their own choices. Nussbaum (2013) coined the question "the people's ability to work and to be?" and expounded on the availability of work opportunities by taking into consideration income at the micro level and GDP per capita at the national level, which provided new dimensions to the policymakers of welfare economics.

Banerjee & Duflo (2005) explained 'why poor countries remain poor' by arguing that the intervention of capital in individual production functions at fixed cost leads to disparity in total production and Total Factor Productivity in developing and developed world. Banerjee & Dufflo (2019) argued that we cannot build a humane world based on laws of economics. To address the problem of poverty, policymakers in government look at how economic growth can alleviate poverty of the poorest of the poor people.

People who live below the poverty line have a very low standard of living. So, measures based on aggregate national income are utilized to see if economic growth has affected the incomes of the poor. This cascading effect on the poor is helpful to improve their living standards. Banerjee and Duflo (2012) have categorically pointed out that the "poor are no less rational than anyone else-quite the contrary". Indeed, as people who live in scarcity, they put much more careful thought into their choices; the question of their survival makes them sophisticated economists. Economic production creates wealth which is distributed as income. Income is used for consumption which results in personal utility on the part of the earner. In economic terms, utility is judged as economic well-being (Gasper, 2000). So, economic well-being is the result of income generated from higher production inputs. Higher input to the process causes more demand for labor, thus resulting in employment. This allows further consumption, satisfaction of personal utility, and well-being of people. Hence, well-being comes from income; which results in poverty reduction. IMF literature noted that to some extent Deaton's studies tried to erase the confusion between evaluative and hedonic well-being through the concept of the Deaton Paradox, which acknowledges that the immediate income shock has small impact on the consumption of individuals (Walker, 2018). He found that various subjective measure has been used in order to give new insight and capture a number of difficult phenomena, such as tradeoff between inflation and unemployment, the cost of air pollution, or the values attached to environmental amenities. Drèze & Sen (2013) advocated the role of education in human development. They discussed and emphasized the contribution of education not only to economic growth but also to demographic change, political democracy, social equality, and many other dimensions of development.

The literature on Human Development has established that human development is about the richness of human life rather than the richness of economies. Human Development is the expansion of individual choices. At the same time, it has certain aims and objectives which make it both a process and an outcome. Human Development infers that individual must participate and influence the processes that shape their lives. In this phenomenon economic growth is a catalyst to achieve human development.

Human development is broader than other approaches such as human resource approach term, the basic need approach and the human welfare approach. Human development is concerned with the overall development of human life by enhancing human capabilities through regular intervention in the process that shape their lives and for the people by civilizing their lives (UNDP, 1990). At the same time society must ensure whether its current development or wellbeing is sustainable or at the cost of future generations. United Nations Brundtland Commission Report conceptualize the Sustainable development as "development that meets the needs of the present generation without compromising the ability of the future generation to meet their own needs" (UN, 1987). This implied socio-economic and environmental aspect for the present and future well-being. The idea of sustainable development blossomed in goals and targets of Sustainable Development Goals (SDGs).

Measurement of Poverty

The idea of human development has penetrated in policy makings and public debate. The idea finds mobility through the annual Human Development Report published by United Nation Development Programme. Since 1990, the report sought to raise the issues concerning the perspective of human development. Human Development Report becomes the yardstick to assess the through the lens of human wellbeing. Countries across the world start developing their own human development reports in addition to 'global' report.

Human Development Index

From the array of human issues human development identified the three fundamental dimension i.e. health, education and standard of living. The report sought to bring the achievement in expansion of human life. The yardstick of measuring development is compiled in Human development Index (HDI). The HDI has changed the locus of measurement of country's development from economic growth to the progress in wellbeing and capabilities of people of the country. The HDI is an index that captures the fluctuation in key dimension of human development, they are …

i. Longevity (a long and healthy life), measured by life expectancy at birth
ii. Education (access to knowledge), examined by mean and expected year of schooling
iii. Income (decent living standard), based of GNI per capita (in USD, PPP terms)

The value of HDI is the geometric mean of three indices. UNDP has explained the methodology to compute HDI in Human Development Report.

Multidimensional Poverty Index

HDI has its own constraints, it associated only to the dimensions what human development involves. Factors like poverty, inequality, empowerment, human security do not resonate sufficiently in HDI.

Bourguignon and Chakravarty (2003) observed poverty as the deprivation of each dimension of human wellbeing on each to identify poor. Amartya Sen, firm believes in multidimension nature of poverty, powerfully advocated to consider multidimensional approach to poverty and development: "*Indeed, it cannot be denied that avoiding pain and suffering must be a good thing, or that happiness is an important reward of living. No ethical accounting can really ignore this elementary understating. But even those who concede this readily may identify many other features of human life and social events that are also significant. Why not take note of them, in*

Table 1.1 Convergence of indicators of HDI and MPI into SDGs

Dimensions	Indicators	SDGs
Health	Nutrition	SDG2 (Zero Hunger)
	Child mortality	Good Health and Wellbeing
Education	Year of schooling	SDG4 (Quality Education)
	School attendance	
Standard of living	Cooking fuel	SDG7 (Affordable and Clean Energy)
	Sanitation	SDG6 (Clean Water and Sanitation)
	Drinking water	
	Electricity	SDG 7 (Affordable and Clean Energy)
	Floor	SDG11 (Sustainable Cities and Communities)
	Asset	SDG1 (No Poverty)

Source Alkire and Santos (2010)

addition to utilities (in the form of happiness, desire fulfilment or whatever metric the utilitarians advocate)?" (Sen, 1979, 2010). Sen's perspective called for an adequate indicator for human poverty. A new method of poverty measurement was developed by Sabina Alkire and James Foster (2011), akin to Sen's (1993) manifestation of poverty as capability deprivation.

UNDP (UNDP, 2010) adopted the Multidimensional Poverty Index (MPI) to measure different facet of poverty in Human Development Report of 2010. The MPI mirrors HDI dimensions with added indicators (See Table 1.1). New indicators were added, the number of indicators increased to ten: addition of two each in health and education, and six in living standard. Individual would be the unit of analysis, through Alkire-Foster's methodology (Alkire & Santos, 2010).

Since 2010, UNDP is releasing the data of MPI for more than 100 countries. The MPI report becomes a tool to examine the improvement against first goal SDG 1 (No poverty) and its spill over impact on others other goals (see Table 1.1).

Millennium Development Goals: First Global Agenda for Human Development

The Millennium Summit of September 2000 adopted the UN Millennium Declaration, for a new global partnership to reduce poverty, with a deadline of 2015 that is known as the Millennium Development Goals. Millennium Development Goals (MDGs) have become the most widely accepted yardstick of development efforts by Governments, donors, and NGOs. The MDGs are a set of numerical and time-bound targets related to key achievements in human development. They include halving poverty alleviation and hunger, achieving universal primary education and gender equality, reducing infant and child-mortality by two-thirds and maternal mortality by three–quarters, reversing the spread of HIV/AIDS and other communicable diseases, and halving the proportion of people without access to safe water.

The MDGs are (UN Millennium Project, 2006):

- Eradicate extreme poverty and hunger
- Achieve universal primary education
- Promote gender equality and empower women
- Reduce child mortality
- Improve Maternal Health
- Combat HIV/AIDS, malaria and other diseases
- Ensure environmental sustainability
- Develop a global partnership for development

Success and Limitation of Millennium Development Goals

In the final report of Millennium Development Goals, the then United Nations Secretary General applauded the success of MDGs, stating: "MDGs are the first attempt at global level to uplift our fellowmen from the abject of poverty, hunger and inequalities" (Moon, 2015). However, while the period of MDGs witnessed the rescue of more than 1 billion people out of extreme poverty, increased children in school, significant improvement in gender parity in primary education, awareness and concern for environmental loss and fostered partnership among the member countries for development, the success is uneven.

Overall poverty reduction is majorly attributed to rapid growth in emerging economies of Asia, including India, China, Vietnam and Indonesia (Wan & Wang, 2019) at the same time in Sub-Saharan Africa's pace of poverty reduction has been too slow or even stagnant, mainly due to persisting inequality, more than 40% population in the region lived under extreme poverty till the end of target year (Schoch & Lanker, 2020; World Bank, 2022a, 2022b, 2022c, 2022d).

Occupancy of women in national parliament increased in sub-Saharan Africa while remaining stagnant in Asia. Four African countries placed themselves in top ten countries in the representation of women in Parliament (IPU, 2024).

Gender gap still persist in the access to available economic, social and political opportunities (Fan & Sturman, 2019; Spiggle, 2021; Sassler & Meyerhofer, 2023). Women receive twenty four percent less money than men (UN Women, 2015). ILO statistics indicate that nearly three fourth of the working age men (72%) were engaged in labor force while at the same time women's share in global labor force is below 50% (ILO, 2022). Despite the same level of qualification, rate of unemployment is higher among women in comparison to men (United Nations, 2015; Hansen & Quintero, 2023).

The gap between developed and developing countries, rich and poor households within country is reflected in healthcare for children. Chances of stunting among the children from the poorest twenty percent in developing world is double in comparison to the developed world. Children are four times more likely to be out of school in poor household against the children in wealthy households. Despite the progress in access to improved sanitation services, by 2012, nearly half of the rural population (47%) availed the improved water and sanitation facilities against 80% in urban area (UNICEF, 2014).

On the environmental front, Asian countries including China, India and Vietnam added 2.2 million hectares in their forest area while the forest area in South America and Africa contracted by 4 and 3.4 million hectares in between 2000–2010 (FAO, 2010; Jong et al., 2021).

An estimated 4 billion people, around two third of the human population are facing severe water crisis at least 1 month in a year. Half of them live in China and India (Mekonnen & Hoekstra, 2016). Poor and vulnerable population whose livelihood depends directly or indirectly are more prone changes in the biosphere.

The inconsistent advancement partly shows a trend over time to shift focus on subset of particular targets that were convenient to implement, monitor and achieve. Notwithstanding, MDG has specific goals for environment (MDG7), the key concept of sustainable development was missing. Environment has the potential to influence the social and economic targets confined in MDG1, MDG2, and MDG6. Gender parity is MDG3 and MDG5, and not discussed throughout the other MDGs. MDG3 and MDG5 confined to measure only equality in primary and secondary education employment and share of women in parliament, maternal mortality and access to reproductive health. Gender equality needs more attention as half the population need better health and equal participation in all existing opportunities.

Moreover, the experience of MDGs has given hope for new set of universal goals to complete the unfinished task of MDGs in a sustainable way. The world leaders recognize the need for sustainable development in order to provide better living condition to all.

Sustainable Development Goals: Second Global Agenda for Human Development

In the mid of the second decade of the twenty-first century, it is apparent that a great deal of success has been achieved in terms of technological advancements, better living standards, and large proportion of people around the globe making a great escape from hunger diseases and poverty, but the reality persists that many are still left behind. World leaders around the globe understood the need for development where no one is left behind and this objective materialized into the Sustainable Development Goals (SDGs) at the United Nations Conference on Sustainable Development, Rio + 20 of the Heads of State and Government and high-level representatives with the full participation of civil society, to renew the commitment to sustainable development. The objective was to produce a set of universally applicable goals that balances the three dimensions of sustainable development: environmental, social, and economic (UN, 2012). Rio + 20 conference of 2012 marked the 20th anniversary of the U.N. Conference on Environment and Development (UNCED) in Rio in 1992 where the participants endorsed "sustainable development" as achieving socio economic development that "meets the needs of the present without compromising the ability of future generations to meet their own needs" (Leggett & Carter, 2012).

Development agenda prior to SDGs are focused on economic while the SDGs are unique and universal framework that inherits many potentially diverging policy goals in socio-economics and environmental sphere while some goals reciprocate on the success of others. The success of agenda depends on whether the host member will be able to channelize such synergy and resolve the prevailing tradeoffs.

At Rio + 20 conference of 2012, Member States agreed to implement a process to develop a set of targets clubbed in goals i.e. Sustainable Development Goals based on the success of Millennium Development Goals. The SDGs replace the Millennium Development Goals (MDGs), which in September 2000 rallied the world around a common 15-year agenda to tackle the indignity of poverty. Table 1.2, depicts the transition of goals from MDGs to SDGs.

SDGs are important milestone for progress and socio-economic development of countries with the pledge to "*leave no one behind by 2030*'. First five goals of SDGs are based on MDG1. Certain SDG goals correspond to the decent living conditions of the people. Some SDGs goals deal with climate change. The goals of SDGs are indivisible and balance the economic, social and environmental dimensions. Prima facie it seems that the attainment of goals depends on distinct policy goals in line with Tinbergen principle which warrant tools of various distinct policies, i.e. synergy between goals. For instance, in India two distinct policies Mahatma Gandhi National Rural Employment Guarantee Scheme providing 100 days job guarantee and Skill India Mission imparting skill to unskilled population collectively reduce poverty (SDG1), eliminate hunger (SDG2), better health and well-being (SDG3), quality education (SDG4), decent jobs (SDG8), and Reduce inequality (SDG10). The success of SDGs depends on the maximum synergy and minimum tradeoff between the goals. Government of India and South Africa has focused their attention to poverty alleviation and zero hunger. Various programs have been launched in both countries to achieve these goals.

Adoption of Sustainable Development Goals

Sustainable Development Goals (SDGs) were adopted on 25th September 2015 by United Nation General assembly and implemented by United Nation Development Program in 2016 (UNGA, 21 October 2015). They were developed on the basis of experiences gained from the implementation and performance of MDGs. The endeavor was to accomplish the objective that could not be achieved under the MDGs. They build on the successes of the MDGs, while including new areas such as climate change, economic inequality, innovation, sustainable consumption, peace and justice, among other priorities. SDGs are officially known as "Transforming our

Table 1.2 Transition of goals from MDGs to SDGs

MDGs to SDGs	
MDG1 (Eradicate Extreme Poverty and Hunger)	SDG1 (No Poverty)
	SDG2 (Zero Hunger)
	SDG8 (Sustainable Economic Growth)
	SDG10 (Reduce inequality)
MDG2 (Achieve Universal Primary Education)	SDG4 (Quality Education)
MDG3 (Promote Gender Equality and Women Empowerment)	SDG5 Gender Equality
MDG4 (Child Mortality)	SDG2 (Zero Hunger, achieve food security and promote sustainable agriculture)
	SDG3 (Ensure health life and promote wellbeing)
MDG5 (Improved Maternal Health)	SDG3 (Ensure well beings)
MDG6 (Combat HIV/AIDS and Communicable Disease)	SDG3 (Ensure well beings)
MDG7 (Ensure Environmental Sustainability)	SDG6 (Ensure Clean Water and Sanitation)
	SDG7 (Ensure affordable and Clean Energy)
	SDG9 (Ensure resilient Industry, Innovation and Infrastructure)
	SDG11(promote sustainable Cities)
	SDG12 (Responsible Consumption and Consumption)
	SDG13 (ensure Climate sustainability)
	SDG14 (Promote sustainable use of Oceans, seas and marine resources)
	SDG15(Protect life on land)
MDG8 (International Collaboration for Development)	SDG16(Promote partnership for the Goals)
Additional	Promote Just, Peaceful and Inclusive Societies

world: the 2030 Agenda for Sustainable Development". Sustainable Development Goals (SDGs) are set of 17 goals and 169 associated targets ranging from hunger eradication, poverty elimination, combating inequality, fighting climate change and reversing the loss of biodiversity and ecosystem, by 2030. The goals are interconnected—often the key to success on one involves tackling issues related with other goals. The SDGs work in the essence of trust and practicality to make the right choices in order to progress life goals in a sustainable way. They make sure that the benefits of development are not just for today but also for future generations. They provide clear procedures and targets for all countries to espouse in harmony with their own main concerns and the environmental challenges of the world at large. The SDGs are a comprehensive schema. They attack the source of poverty and unite the world together to make a positive change for both people and planet. UNDP offers support to governments to assimilate the SDGs into their national development plans and policies. Attaining the SDGs necessitates the partnership of governments, private sector, civil society and citizens (UNDP, 2017).

Aspiration and Progress of SDGs

Sustainable goals reaffirm the pledge of human development, targets adopted the sustainable path to free the world from poverty, hunger, inequality, ensure access to quality education and health services, protect the biosphere and galvanize the global partnership to achieve the goals by 2030. The aspirational goals need active patriation of all the sectors (state and non-state actors) and each human beings to achieve the targets by 2030.

Year 2022, is the midpoint year to achieve the Sustainable Development Goals. Sever year journey marked progress in forging collaboration between global organization to push the goals forward (Grainger-Brown & Malekpour, 2019). Leading industries responded positively to dedicated targets of economic growth through sustainable industrialization and sustainable production (Agarwal, Gneiting, & Mhlanga, 2017). Strategies of sustainability and approach to sustainable development occupied their space in institute of higher education, through tested model of 'education for sustain development' (UNESCO, 2017). The whole concept is to empower individual to become agent of sustainable development. Member states have prioritized the agenda of global goals in their national policies, for instance India, South Africa, Ethiopia, Indonesia Nepal, Rwanda and many more (OECD, 2017). However, the progress of SDGs was halted by COVID-19 pandemic. The success of global goals not only halted but start reversing for continuous three year. After continuous toll of pandemic, Russia-Ukraine conflict make the situation worse for humanity. Consequently, there is a significant increase in evidences of poverty and hunger, income constrained translated in to health deprivation and deacceleration in access to education (United Nations, 2021).

Sustainable Development Goals and Targets

This book discusses the progress made by India and South Africa in path of SDGs in chapter II and Chapter III respectively. Here the book attempted to present the global progress of each goal in respect to India and South Africa.

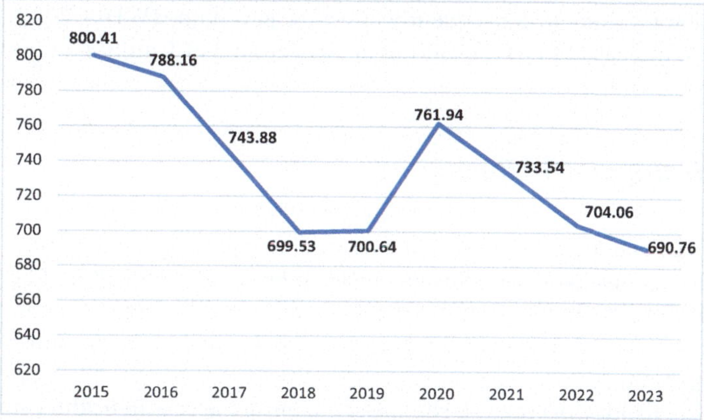

Fig. 1.1 Global Poverty trends (below $ 2.15), from 2015–2023. *Source* World Bank (2023)

SDG1: No Poverty

According to World Bank statistics, the incidence of global poverty deaccelerated 9.13% from 10.81% during 2015–2018 (Yonzan et al., 2023). In absolute number, number of persons living below the line of $2.15 a day reduced to 699.53 million from 800.41 million over same period. The progression starts reversing since 2019 and picked up during 2019, the proportion increased from 9.05% to 9.74% between 2019 and 2020. Almost 61 million additions in global number of poor. The deterioration in poverty incidence reverses the achievement of more than four years, the recent data shows that in 2023, the proportion returned to the pre-pandemic level (see Fig. 1.1).

The data indicates that global poverty is declining and descended closer to the pre-pandemic level. It also revealed that the world has lost three years in the fight against poverty, driving the prospect further from meeting the goal of poverty eradication by 2030.

SDG2: Zero Hunger

The target to achieve huger free and food secure world is on off track. The number hunger and food insecure persons keep rising since 2014. FAO report revealed that progress on hunger becomes stagnant since 2015. Incidence of undernourishment rose to 9.8% in 2021 from 8.0% in 2019. Further, report depicts that around 702 and 828 million people across the world faced hunger situation in 2021. An estimated addition of 150 million people since 2019 (FAO, WTO, World Bank, 2023).

World Health Organizations statistics shows that prevalence of anaemia among women (age 14–19) has progressively risen, the number reached to 570.8 million in 2019 from 493 million 2000, potential to increase the cases of mortality and morbidity among women, consequently affect the new borne (WHO, 2023). Number obese people increased to 675.7 million in 2016 from 343.1 million in 2000.

SDG3: Health and Wellbeing

SDG3 ensures better health and wellbeing of all. In pre pandemic period, progress was recorded in health sector. Advancement in key health indicators like maternal and child health care, infant mortality, immunization and limiting the communicable diseases was recorded, however the progress was inequal between rural and urban and rich and poor.

Universal Health Coverage (UHC) is key to ensure access to health care services required by the persons without any financial constraints. It is seminal to realize the health goals and driving force behind health and nutrition targets, subsequently accelerate the pace of poverty reduction.

A joint report of WHO and World Bank (2023) noted that despite significant improvements in the last two decades, further advancement in UHC was stagnant during 2019 and 2021. The global average of the UHC Service Coverage Index (SCI) reached 68 in 2019 from 64 in 2015, and the progress stalled since 2019. Healthcare services consume more than 10% of the household budget of poor families, and the share of out-of-pocket spending has continuously increased from 9.6% to 13.5 during 2000 and 2019, over the same period the count of people increased by 76%, from 588 million people to 1.04 billion people.

The pandemic has devasted the health of individuals and subsequently the health infrastructure. An estimated more than 500 million peoples are infected by the corona virus leading to death of 15 million persons directly or indirectly, by the end of 2021.

Global trends show that, proportion of deliveries attended by skilled health workers, doctors, nurses and midwifes increased to 84% during 2015–2021 from 77% in 200 there was an increase in 2008–2014. Under 5 mortality and neonatal mortality reflects tepid growth. Under5 mortality slipped to 37 deaths per 1000 live birth in 2020 from 43 deaths per 1000 live birth, in 2015. The HIV incidence recorded 39% deceleration between 2010–2020 against the 75% target, agreed by United Nation in 2016. Still, approximated 1.5 million people were living with HIV infection which leads to 680,000 deaths in 202. Despite progress, the target of halting malaria incidence was off from target set by WHO, in 2020 rate was 35 cases per 1000 persons against 59 cases per 1000 person's target. Pandemic and its containment measures halted the infant immunization process, coverage reduced to 83% in 2020 from 83% 2019 (United Nations, 2022a, 2022b, 2022c).

SDG4: Quality Education

Target of SDG4 is to keep pupils in school and ensure completion of their education. Despite significant improvement in education goals in pre- pandemic period, but the progress is not sufficient to meet 2030 education goal. In spite of reduction in share of children and youth out of primary and secondary education from 19% in 2010 to 17% in 2018, an estimated 258 million child and youth were out of education institutions till 2018 (UIS, 2019; United Nations,2022a, 2022b, 2022c).

Despite the increase in access to education over the year, persistent inequalities among the various groups of the societies are bigger threat for educational goals. The children of poor families are consistently more deprived in educational opportunities and outcomes than children in urban and rich cohort. The basic school infrastructure, access to clean and drinking water, electricity, and toilets are far from universal. Global trends shows that 25% primary Scholls lacks basic amenities and 50% are deprived from computer and access to internet services.

Many families rely on them to provide daily meals for their children and childcare while parents are at work. In this note we focus on a different cost. COVID-19 containment measure has intensified the slow progress in SDG4. Almost all countries have closed the school as containment measure. at the pick of Covid-19, nearly 94% of pupils were deprived from classes in school (World Bank, 2020). Many countries switched to distance learning as remedy to school closure, but due to unequal access, 500 million pupils predominantly in global south are deprived from education (United Nations, 2022a, 2022b, 2022c; UNESCO, 2023). In long run, disruption caused by pandemic adverse influence the productivity and contained the social and behavioral progress.

SDG5: Gender Equality

Gender equality and women's empowerment are vital to humanity. The spill over impact of gender equality has potential to accelerate each goal of SDGs. It is manifested that, dreaming of any development goal, could not be realize without the full participation of half population. Protection of women's right across the 17 goals is the only way to realize the dream of sustainable development.

Trends reflects that the current pace is not sufficient to achieve the gender goals by 2030. Despite the progress, the world is trailing behind the targets in core areas like, time spent in care economy, decision making in respect to sexual and reproductive health, and gender budgeting. Health facilities for women are poorly funded. Women are deprived in every country and culture, so as the violence against women persist across the classes of society.

WHO (2021) data shows that women aged 15 year and above who have ever partnered, experienced physical or sexual violence of at least for one time their life scale by intimate partner or husband. Such violence in women aged 50 and above is underestimated because of data limitation. The data shows that 4–7% of women in this age group subjected to physical and sexual violence in last 12 month during 200–2018.

Despite the advancement in international standard and awareness regarding child's right, child marriage is global concern (Akçay et al., 2024). Last five-year data shows 10% deceleration in global rate of child marriage. However, still the proportion of young women marriage (under 18 year) is 1:5. Sub-Saharan Africa and South Asian countries are hot bed of child marriage where 37% and 30% women under 18 years respectively were weeded in childhood (Subramanee et al., 2022). Trends shows 100 million child brides would add by 2030, pandemic has added more 10 million in previous projection (UNICEF, 2021a, 2021b).

Role of women in decision making is widely acknowledged, while the rate of progress in increasing the proportion of women representation is disheartening. In between 2015–2022, the share of women in national parliament ascended to 26.2% in 2022 from 22.4% in 2015. However, occupancy of women in third tiers is more than one-third.

In order to achieve gender equality, it is important to protect the right of women through legal provision and polices. Right to land and right to property are vital in reducing the dependency of women on men, unfortunately merely 15 out of 52 countries have such provision to protect the right of women in this regard.

Public finance is fuel to gender equality. The available data shows that, during 2018–21, out of 105 countries, 26% of countries have gender responsive budget, another 59% have some elements of gender equality while 15% countries are not showing any features of allocation for women advancement.

The tepid progress, in expanding women's right and opportunities, scaling from economic participation and decision making, is ceased by pandemic. Working women are more prone to unemployment and exclusion from job market. COVID-19 has intensified the dropout ratio and increased the number of women not in education not in training. Women and girls are facing discrimination numerously, especially the disable girls and women. UN Women's (2022) literature estimates that with the current pace, it would take 286 years to bridge the gap in legal protection, 140 years to occupy equal representation in managerial and leadership position at the work place and at least another 40 years to realize the target of gender equality in national parliament.

SDG6: Clean Water and Sanitation

Water is key to sustain humanity and vital to SDGs. Robust economic growth along with surge in population and quest for urbanization has raised the demand of water resources, consequently, 36% of the global population are prone to water scarcity. Depleting water sources posing threat to sustainable development and challenge to human development.

The world Bank data shows that, an estimated 2 billion people deprived from basic sanitation services and around 673 million people are openly defecate. The sanitation crisis is uneven distributed across the societies. In rural cohort, 91% people practice open defection and 72% lacks basic sanitation services. Water crisis has differential gender impact, it exuberates the health issue of women and girls in their pubic age.

However, proportion of population access to clean drinking water facilities rose to 73% in 2022 from 69% in 2015, over the same period proportion of people's access to adequate sanitation services increased to 57% from 49%, still 2.2 billion persons are still deprived from access to managed water and around 3.4 billion persons lack adequate sanitation services (UNICEF & WHO, 2023). Projection shows that, with same tempo, by 2030, world will achieve only 81% of targets related to clean water and sanitation. The Frequent hand wash is key to limiting communicable diseases like COVID-19 and other transmittable diseases. During 2015–2020, the proportion of people who used soap and water for handwashing at home increased from 67 to 71%, still, 1 out of 4 people cannot afford soap and clean water for hand washing. WHO (2022) report, suggest that double pace of progress is required to achieve the universal coverage of clean drinking water and four-fold speed is needed to universal sanitation services to achieve the targets 2030.

SDG7: Affordable and Clean Energy

Clean, safe and affordable energy is core to sustainable development. Energy is fuel to economic development across the world. As the world population growing, the demand of energy is on continuous surge, since it is a key input to agriculture, transport and communication, industries and commerce, and other necessities to improve the standard of living.

Increased in energy demand has its own consequence. Supply and management of energy create environmental pressure. Thus, mitigating the negative impact by accelerating the production of affordable, clean and sustainable energy is need of time. Global progress has registered in advancing the targets of sustainable energy. However, current stride is disappointing and improvement is insufficient to meet the SDG7 by 2030.

Access to electricity reached to 91% in 2020 from 83% in 2010 (IEA, IRENA, UNSD, World Bank & WHO, 2022). The progress leads to reduction in number of people living with electricity to 733 million from 1.2 billion. Weak improvement in safe and clean cooking fuel has serious impact on health issues, it leaves 2.4 billion prone to air borne diseases. Although, progress is uneven, marginalized and vulnerable are more deprived than the people live in urban areas and belongs to wealthy households.

In recent year, the rate of access to electricity fell from 0.8% age point during 2010–2018 to 0.5% during 2018–2020. Slow progress is further intensified by COVID-19, mounting input cost (commodity, energy and transportation cost) raised the production and shipping cost of solar cells, wind turbine and biofuels are other reasons to doubt the attainment of affordable energy goal by 2030.

SDG8: Decent Work and Economic Growth

Decent work is fundamental to human development, so as to sustainable development. Decent work includes opportunities for productive job with fair income along with social safety benefits and security at work place, avenue for social integration, free to express their issues, organize and participate to become the active agent of democratic process which influence their live and improve the chances for equality of opportunity without any discrimination based on gender, caste, creed and religion.

Further, it contributes in government treasury by increasing the number of tax payers. Subsequently, empower government to increase the spending in social sector. Thus, strong social sector produces more human capital which the standard of living by taking people out of poverty and hunger.

Economic growth is raveled with loss of natural biosphere, thus, limiting the capabilities of societies to cater their future needs. SDG8 promotes idea that economic growth must be unraveled from nature's loss, the economy can expand by taking care of natural resources and mitigating the environmental challenges.

Globally, average real GDP growth rate was increased by 1.6% during 2010–2014, the rate was lower than the rate achieved during 2000–2004, because of North Atlantic Financial Crisis (NAFC) of 2008, in 2018 real per capita GDP growth reached to 2%, the same rate was reflected in annual GDP growth rate over the period 2010–2018. The growth rate decelerates to 1.5% in 2019. Further, in 2020, world has experienced the most devastating situation after great depression, the pandemic, growth rate dropped sharply by 5.6% in 2020, recovered in 2021 by estimated 9.1%. Further, for 2022 and 2023, GDP annual growth was 3% and 3.5% (IMF, 2024).

NAFC raised the global unemployment rate to 6.6% in 2009, which dropped to 6.1% in 2015, further, decelerated to 5.5% in 2019. Due to COVID-19, the global unemployment level increased to 6.9 in 2020, which further decreased to 5.8%, which was still higher than the pre-pandemic level (ILO, 2023). As the world is recovering from the pandemic shock, a surge in informal job market was recorded. However, rising child labor is a major concern, 8.4 million children were added in child labor over the period 2016–2020 (UNICEF, 2021a, 2021b). Share of youth not in employment, education and training rose to 23.3% in 2020 from 21.8% during 205–19.

SDG9: Industry, Innovation and Infrastructure

Sustainability relies on sustainable industrial growth. The way for sustainable future goes through green infrastructure, modernizing existing infrastructure and harnessing the potential of advance technologies can stimulate the negative impact of exogenous factor of production along with generate resilience and boost the effectiveness of natural resources. Heterogeneous economies with strong infrastructure for road and transport, internet connectivity and other services has better capacity to faster recovery. COVID-19 pandemic has manifested the role of modern industry, innovation, technology and resilient infrastructure rebounding the economy and realizing the SDGs.

Manufacturing is engine to economic growth. Ratio of Global manufacturing value added (MVA) in total GDP increased to 16.9 in 2021 from 16.2 in 2015 (United Nations, 2022a, 2022b, 2022c). Global manufacturing output surpassed the pre pandemic level in 2021 to 7.2% from dropping 1.3% during 2020. However, share of job in manufacturing was declined in 2020 to 13.1% from 13.7 in 2019. Over the period 2015–2021, penetration of 4G connection doubled, covering 88% of global population. In 2021, mobile broadband connectivity reached to 95% global population. However, 17% population of least develop countries and land locked developing countries are deprived from broadband coverage. Thus, least develop countries and land locked developing countries lags behind the universal target of affordable internet connectivity.

SDG10: Reduce Inequalities

Policy makers across the world are trying effort to minimize the inequality. All human beings are born equal, deprivation and impoverishment have differential impact on capability building. Income plays a major role in translating capabilities in functioning. Thus, lack of income restricts the upward mobility of individual in society and economy. Persistence inequality monetary as well as social within and between geographies is major threat to human development.

Global goals aspire to reduce inequality by raising the income by increasing the income growth of lower 40% faster than national average by 2030. The increased income has potential to avail decent life and subsequently reversing the trends of inequality. World Bank noted that before pandemic, income inequality was contracting within countries and Emerging Market and Developing Economies (World Bank, 2022a, 2022b, 2022c, 2022d). Trends in many countries shows the declining trends in inequality, income of poor rose faster than national average, however, same pace was missing in social inequality.

Pandemic has reversed the achievement of inequality target. People having low income are more prone to lags behind. Pandemic has exacerbated the existing inequalities. Slow recovery and structural heterogeneity expand the inequality gap within and among countries.

Absolute number of refugees was historically high in 2021, further, Russia-Ukraine war force more people to move from their place, thus, world has witnessed biggest refuge crisis in recent times (UNHCR, 2022).

SDG11: Sustainable Cities and Communities

Urbanization is closely associated with economic growth. Higher level of development mirrored in urban settlement. Cities are major contributor of GDP, 80% of global GDP related to urban settlement. City provide avenue for innovation and growth required to accelerate sustainable development.

City provide avenue for innovation and growth required to accelerate sustainable development. Globally, urban settlement accommodates more than half population. Trends shows that, by 2050, an approximate 7 people out of 10 people would belongs to cities. Rapid urbanization has serious impact on natural ecosystem. Cities and town are experiencing the catastrophise natural disasters such as flooding, storm and drinking water crisis. Cities located along coastline are on verge of drowning.

However, urbanization has its own flaws, 25% of urban population live in slum. They are deprived from basic sanitation facilities and clean drinking water; lack of hygiene is major challenge. COVID has exposed the pathetic living condition of slum dwellers. Studies shows that 1% increase in urban population leads to 2.3% and 5.3% increase in population of slum dwellers in Asia and Africa (Shulla & Kőszeghy, 2021). To walk out from poverty, nearly 1 billion slum dwellers are waiting for support from national and international for their mainstreaming and thus, realization of SDGs.

SDG12: Responsible Consumption and Production

Sustainable growth as well as sustainable development can realize only by limiting the toxic and pollutant generated throughout process of production and consumption. Sustainable consumption is core to sustainable development. Irresponsible consumption pattern is call for depletion of natural resources, climate change and weakening of ecosystem. Further, at some extent, these crises can halt the progress of human development. SDG 12 promulgate the idea of sustainability in production and consumption.

Domestic material consumption was ascended by 65% during 2000–2019. Over exploitation of natural resource intensified the pressure on natural resources. Easing pressure can be done by increasing efficiency and effectiveness of existing resources. at the time when countries are attempting to become food secure, food wastage is consistently raising across the world. In 2019, early 17% of total produced food grains was never reached retail market (UNEP, 2021). Annual loss 931 million tons of total food translate in to 121 kg per capita loss annually.

SDG13: Climate Change

Climate change is the primary challenge in meeting the sustainable development. It impedes the progress of human development and push people in trap of never-ending poverty (Kyte, 2014). Studies suggest that policy makers can mitigate the welfare cost by addressing the challenges of climate change (Gomez-Echeverri, 2018).

Unarguably, anthropogenic activities have unprecedently raising the temperature of globe. Anthropocene in industrial raised the global temperature by nearly 1.00 C, if current rate sustain, project range would be 1.5 °C–2 °C across the period 2030–2052. Hottest period in recent years recorded in 2011–2012, resulted in sea ice melting to historical low level. Consequently, rise in surface temperature threating the coral reefs. To meet the challenges of climate change, 175 countries had signed the Paris agreement in April 2016, with objective to limit the future carbon emission, and resource mobilization for investment in policies needed to realize sustainable development. Developed world has committed to mobilize $100 annually by 2020, further expanded to climate action programme in developing world. While IPCC estimation suggest $1.6 trillion to $3.8 trillion annually required to contain the global temperature below 1.5 °C.

SDG14: Life Below Sea

Healthy seas, oceans and marine ecosystem are core to blue economy, and vital for human health and wellbeing. Healthy seas, oceans and marine ecosystem are core to blue economy, and vital for human health and wellbeing. Oceans draws nearly 23% of carbon mission and 90% of heats generated by human activities (Castano-Isaza & Diez, 2021). World needs a cohesive plan to leverage the strategies adopted by countries to deal with climate change, sustainable use of seas, oceans and marine cost for sustainable development envisaged in SDG14, life beneath sea.

Marine resource plays major role people lives in coastal areas by providing food and income. However, if these are ill managed, they hinder the path of human development. Fish and other marine animals are primary for global food security and crucial for nutrition intake. The sector is under persistent threat due unregulated, illegal and unreported incidence of fishing. However, with minor fluctuation, the production of fisheries is stable, substantiating the average catch over the period 1995 to 2020. In 2020, production of aquatic animal for human consumption reached to 157 million tons (FAO, 2022).

Another major threat to aquatic life is plastic, nearly 80% of lotters are plastic, an estimated 800 species are under threat by marine debris (United Nations, 2016). Studies shows that 13 million metric tons of plastic dumbed in ocean every year (Reddy, 2018). Plastic threat is not limited to aquatic animals, we humans are equally prone to danger, as it enters our bodies through sea food we consume.

SDG15: Life on Land

Conservation of biodiversity is vital for sustainable development goals. SDG15 call for sustaining life on land, by making coherence between human and finite nature. Nature has tendency to revitalize the harm caused by human activities. Although, unmatching population growth, rapid industrialization, flourishing urbanization and expansion of cultivation all are taking place on finite landscape leads to contraction of landscape and degrading of ecology. Due to anthropogenic activities, almost 40,000 species are on brink of extinction, on an average 10 million hectares of forest are encroached annually, and still nearly half of the existing biodiversity are not protected. Studies suggest that humanity is experiencing the sixth major extinction in human history (Bradshaw et al., 2021).

Globally, forest area is continuing to shrink, the share of forest remains to 31.2% in 2020 which was 31.9% in 2000, loss of 100 million hectare, however the pace of contraction slower than previous decades. Nearly 90% of deforestation was resultant of agricultural expansion (United Nations, 2000). Countries across the world are prioritising ecosystem and biodiversity in their national policies to meet the challenges of deforestation. Neglect of biodiversity will lead to unstable future.

SDG16: Peace Justice and Strong Institutions

Peace, conflict and Justice are core to global goals. Peace and development occupy central stage in any debate discussion related to humankind. Peace catalyzes development or vice versa, the situation mirrored in conflict zone, chaos and uncertainty dampen the prospect of development. In recent decades worlds has been experiencing more peace and more secure. However, prevalence of peace and security is uneven within and between the countries. However, after all advancement, vulnerable, marginalized and children are still not able to access institution meant for justice and peace.

Onset of COVID-19 crisis, 100 million people were dislocating from their home, further Russia-Ukraine war has intensified the displacement forced displacement. The toll of war and conflicts has been unbearable poor and vulnerable leading to increase in number of cases related to human right violation. For achievement of aspirational global goals, there is need of efficient and effective institution to protect the human right of every individual and provide a better world by 2030.

SDG17: Partnership for Development

Partnership is pre-condition of sustainable development goals. While all the goals are interlinked and framed to accelerate the progress of human development, conducive to new world by 2030, by acknowledging challenges of people, planet and prosperity. Weak infrastructure in developing world impede the progress of human development. High inflation after pandemic and Russia-Ukraine conflict decelerates the progress despite the strong inflow of Official Development Assistance (ODA) and Foreign Direct Investment (FDI).

To take world on path of SDGs, there is a need of restructuring of debt and other international financial management, at the time when world is stuck destitute, uncertainty and conflict.

Uneven Progress

Global trends show the successful transition of global goal of humanity; from MDGs to SDGs and further examine the global progress and negative impact of COVID-19 and ongoing Russia-Ukraine conflict in mid-way to SDGs. We found the progress uneven within and among the countries. The unequal development is more glaring in country with heterogenous demography and geography.

Thus, our study proceeds with examination of two countries from two different continent with similarity in diversity, India and South Africa. The global trend of MDGs success is reflected in the progress of India and south Africa. Both countries have achieved significant progress, but the improvement did not translate in the success of all MDGs (see Table 1.3). India has achieved MDG3 and MDG6 completely. India has registered significant progress in reducing the proportion of poor, achieving gender parity in primary education and secondary education, halting and reversing the spread of communicable diseases, succeeded in clean water supply and conserving the environment. Though, targets of eliminating hunger, deaccelerating maternal mortality rate and ensuring sanitation services shows weak performance (MOSPI, 2017). South Africa has significantly reduced the number of poor, improving the access of basic services like education, healthcare, education and social assistance to historically marginalized class. It has partially achieved the targets of MDG2, MDG6, MDG7 and MDG8 and almost succeeded in realizing MDG3, while the progress in MDG1, MDG2, MDG4, MDG5 were not as expected (STATS SA, 2016).

The COVID-19 Pandemic

The COVID-19 pandemic has put the world economy on standby. It is a threat to global public health caused by sudden emergence of novel coronavirus (2019-nCOV; Jin et al., 2020); (Zhu et al., 2020). World's first case of novel corona virus was reported in Wuhan, China in December 2020 (Gautam & Hens, 2020). Within a month the corona virus swept through the globe. On 13th January 2020, Thailand emerged as first country after China to detect coronavirus patient. On 15 January

Table 1.3 Report card of MDGs in India and South Africa

	India	South Africa
MDG1	Partially achieved	Not achieved
MDG2	Partially achieved	Partially achieved
MDG3	Achieved	Almost achieved
MDG4	Partially achieved	Not achieved
MDG5	Partially achieved	Not achieved
MDG6	Achieved	Partially achieved
MDG7	Partially achieved	Partially achieved
MDG8	Partially achieved	Partially achieved

Source MOSPI (2017) and STATSSA (2016)

2020, Japan reported the first case of corona virus and five days later, on 20th January South Korea confirmed its first patient of coronavirus infection. On 24 January 2020, France became first European nation to report confirmed case of coronavirus (WHO, 2020a, 2020b, 2020c; Indian Express, 2021). In India, the first case of corona virus was reported on 27th January 2020 when a 20-year-old lady with travel history of China was admitted in emergency ward of General Hospital, Thrissur, Kerala with symptoms of sore throat and dry cough (Andrews et al., 2020). On 30th January 2020, World Health Organization (WHO) declared global emergency of international concern as the count of confirmed cases reached to 7711, and number of suspects rose to 12,167. By then, 170 people had died and 1370 were critical among the confirmed cases in China (WHO, 2020a, 2020b, 2020c). Additional, 7500 confirmed cases were reported from 20 different countries (Khan et al., 2020).

Eventually, the horrifying news of the rapid spread of the contagious COVID-19 started coming from every corner of the world. Elwen Callaway's (2020) study identified the rapid expansion of coronavirus across the world and qualified the criteria to be declared a Pandemic. On 11 March 2020, World health Organization (WHO) characterized COVID-19 as a pandemic (WHO, 2020a, 2020b, 2020c). In the beginning governments were aware about the contagious nature of COVID-19 but they were mostly unequipped to immune their people from the virus. The major hurdle was lack of any effective medicine or vaccine to neutralize the situation (Rolland, 2020). Unchecked, the virus freely travelled across the world by extensive movement of human beings. Studies of Lau and et al., (2020), Remuzzi and Remuzzi (2020) finds that strict confinement of people in high-risk zones have potential to reduce the pace of COVID-19 infection. The study was based on the lockdown experience of Italy and China. Further, lockdown as a containment measure was instituted in many countries to break the chain of infection. Social distancing, vigilant sanitation and hand wash, home quarantine, facemasks, prohibiting public gathering and means of transportation were measures instituted as countries came under lockdown. The lockdown has in turn had a disproportionate impact on developed and developing countries.

In response to the swift spread of novel corona virus, countries of Asia and Africa imposed stringent lockdown as a measure to protect the people from any further infection. The lockdown also helped government to strengthen public health system in order to cater to the unprecedented health calamities. However, lockdown has had differential impact on the rich and the poor, daily wagers and salaried person, young and aged people and also on male and female. For affluents people living in the metropolitan cities, lock down entailed shift to over-the-top (OTT) from their regular jobs, but for the daily wagers and relatively poorer and marginalized section of the society, lockdown brought uncertainty, loss of jobs and rise in hunger. The differential impact of the pandemic is also evident when we take into consideration of the fact that the larger proportion of people in developing countries are employed in informal sectors. Aged people, especially with underline medical condition are more prone to infection than youths.

Harsh policing in developing countries also contributed in creating an atmosphere of fear and uncertainty. A person living on the margins was more likely to be welcomed with sticks in making any attempts to go outside to buy essential commodities. Most of the countries in the developing worlds are densely populated and therefore maintaining social distancing was a huge task. And yet, government choose policing to maintain social distancing.

Rationale and Relevance of Book

Sustainable Development Goals (SDGs) were adopted by the United Nations to complete the unfinished work of Millenium Development Goals (MDGs). While countries around the globe have achieved substantial progress towards attaining SDGs, the COVID-19 pandemic was an unexpected disaster that halted and even reversed the progress achieved in terms of human development. The pandemic laid bare the uneven development among the countries as well as within the country. Countries in Africa and South Asia are home to the world's most deprived population for instance, India has emerged as the world's most populated country but data indicate that half of India's youth are not employable. The second most industrialized country of Africa, South Africa is the most unequal country in the world with nearly half of its working age population unemployed. In the last two decades, India has shown remarkable progress in economic growth and in the post covid era India has emerged as the fastest growing economy, but at the same time the country is not performing well in the hierarchy of hunger index, human development index, happiness index and gender party index. On other side, despite the sluggish growth rate, South Africa performed better than India. This mismatch needs to be addressed.

This book investigates the progress made by India and South Africa in achieving SDGs. The study deciphers the gaps that exist in the context of India and South Africa in transition from MDGs to SDGs. Collectively, it attempts to explore the strategies that should be adopted by India and South Africa in order to achieve successful transition from MDGs to SDGs and identify the challenges and shortcoming faced by India and South Africa in the implementation of MDGs to SDGs. Finally, to provide a way forward by inferring from the expertise of each other and analyzing the potential of cooperation between India and South Africa to achieve the goals laid down in MDGs and SDGs.

Literature Review

Successes of Sustainable Development Goals will rely on the experience of Millennium Development. SDGs are interlinked like MDGs where success of one goal accelerate the progress of others goals and the pathway of SDGs will be identified and achieved by global problem-solving network provided by new social media and information technology (Sachs, 2012). Technological advancement is a key to human prosperity, as it can exert significant impact on social, economic and environmental pillars of SDGs (Huang, 2021).

Much of the existing literature focuses on the idea of interlinkages between the 17 goals. The study of SDGs begins with emphasis on better understanding between the targets, synergy between technological advancement and its global consequence (Lu et al., 2015). SDGs have the potential to provide for the future of humankind. It holds all the factors that contribute towards a sustainable future (Costanza et al., 2016) and the progress of SDGs has been achieved by incorporating the ideas and expertise from various experts, partners and stakeholders (Rickels et al., 2016). The SDGs agenda will be able to deliver on its potential if mutually fortifying actions are taken with minimized trade-off (Nilsson et al., 2016).

To achieve the SDGs there should be proper measurement of synergies and tradeoff within and across the goals (Renaud et al., 2022; Min Cao et al., 2023). SDG1 (no poverty) has strong synergy with other goals whereas as SDG 12 (responsible consumption and production) identified as common tradeoff (Pradhan et al., 2017; Kroll et al., 2019). Irresponsible production and consumption will stand in the way of Sustainable Development Goals (Gasper et al., 2019). It has also been shown that policy coherence is essential in the implementation of SDGs, since the goals and their targets are inseparable (Breuer et al., 2019). SDG 3 (health and well-beings) is core to SDGs, it is recognized as essential element in accomplishing all three dimensions of SDGs (Hill et al., 2014; Nunes et al., 2016). SDG 4 (quality education) is crucial for achieving SDGs (Camilleri & Camilleri, 2019), it has synergetic relationship with many goals (UNESCO, 2015).

Studies suggest that good governance must incorporate agriculture and livestock as part of land-use and rural development in synergy with developing bioenergy (Smith & Olesen, 2010), that Climate change and poverty alleviation are inextricably linked (Mathy & Blanchard, 2016), since impacts of climate change will reduce the pace of poverty reduction (Bue & Klasen, 2013). SDG 13 (climate action) has potential to amplify global food insecurity (SDG2), reduce access of clean drinking water (SDG 6), deteriorate people's health (SDG 3), limiting their present and future income (SDG1) (World Bank, 2022a, 2022b, 2022c, 2022d; Filho et al., 2023).

Lusseau and Mancini (2019) advocated similar finding that regulating climate change, contracting inequalities and responsible consumption are the major challenges in the way of SDGs in many countries. On other hand poverty reduction and lowering inequality will have overall impact on SDGs. Francesco Fuso Nerini and et al., (2018) established positive relation between SDG7 (sustainable energy) and others SDGs; Scherer et al. (2019), analyzed the interaction between social (SDG1and SDG 10) and environmental goals (SDG6, SDG 13 and SDG 15); Kroll et al., (2019) recognized that the synergy between certain goals especially SDG1, SDG7, SDG9 rooted in scientific innovation while SDG11, SDG13, SDG16, SDG17 will continue to have trade-off.

Scholarship of Justyna Bandola and Sotiria Greg's (2022) engages with the implicit and explicit ideas behind the framing of global goals. They contend that the success of Millennium Development Goals depends on the donor country while Sustainable Development Goals can be achieved by the localization of goals. Stephanie Marie Seferian's (2021) book proposed a solution for minimalism to sustainable development. Industry relies on natural resources for the production of goods and since resources are finite, over-exploitation leads to an uncertain world for our future generation. Markus Kaltenborn (2020) edited book contents that sustainable development goals can be archived by ensuring human rights of the people. Keitsch and Walter Vermeulen's (2020), edited volume Trans disciplinarity for Sustainability: Aligning Diverse Practices, emphasized on the association of academic and non-academic actors in the production of knowledge to leverage sustainable development goals. Nazrul Islam's (2020) work Rivers and Sustainable Development: Alternative Approach and their Implication covers the looming water crisis with increasing human population and the distortion of the earth's hydrological cycle with respect to rivers. Prachitha et al. (2019) research article pointed that partial achievement of MDGs in India is due to lackluster policy implementation at national level and sub national level. Lomazzi et al. (2014) article figured that set of targets under MDGs are concerned more with maternal health, child health and communicable diseases while less attention has been given to the environmental sustainability and forging global partnerships. Further, the study expected to incorporate the issues of environment in post-2015 goals, an inclusive approach built on coherence between the goals. Walker and Pekmezovic's (2019) book Sustainable Development Goals: Aligning Diverse Practices shows the causality between technology and human development. Further, they added the need for law reform and mobilization of finance as prerequisites of sustainable development in their book Sustainable Development Goals: Harnessing Business to Achieve the SDGs through Finance, Technology, and Law Reform. Swingler and et al., (2012) in their work compare the health status of children born in South Africa and other middle-income countries. The study finding suggests that the status of child in South Africa is in poor condition in comparison to others. The study calls for an urgent need to meet the challenges of health care system by giving special attention to marginalized groups. McMichael et al. (2008) discuss the negative impact of increasing human pressure on environment, distorting the biophysical and ecological system. They contend that there is no way to separate health from environment especially in analysis of developmental agenda. Poverty in developing countries cannot be eliminated without the protection of forest as a large section of people in the developing world depend on forests for their livelihood. Untreated industrial waste contaminates the water used by poor for drinking purpose. Over exploitation of chemical and fertilizer reduces the soil fertility. Thus, the only way to realize the global goals, is the inclusive approach of development that is the sustainable way. Martíneza et al. (2008) research tried to explain the circumstances pushing an individual to vulnerability. The study confirmed the improvement in access to clean water and access to better sanitation services and improved durable structures for slum dwellers. All this improvement leads to deaccelerate under five mortalities, but the chances of survival of under five children in urban household is 5 times more than the under-five child in slum area. Further, study concluded that filling the gap between urban households and slum dwellers would be key to achieve sustainable development. Lagarde et al. (2007) examine the progress in access to health services by infusing conditional cash transfer in developing countries. The study suggests deep assessment of conditional cash transfer scheme in leveraging the health condition of people living in middle and low-income countries. Rehfuess et al. (2006) work shows that most of the households in developing world use the fossil fuel for combustion purposes. On an average, 52% households in WHO regions use solid fuels, the proportion reached to 72% in Sub-Saharan Africa while the consumption of traditional fuels in developed countries is less than 5%. Gundry et al. (2004) article explored the causality between water contamination and prevalence of disease. The study focuses on the influence of drinking water quality at point of uses and their relationship with general diarrhea and cholera. Ravallion (2004) critically engages with the definition of pro-poor growth as any increase in GDP that reduces poverty. He stresses that such a definition is too broad implying that most real-world instances of growth are pro-poor, even if poverty decreases only slightly and income distribution worsens during a period of strong growth. A more appropriate definition has growth as pro-poor in addition to reducing poverty is the reduction of inequality. Despite being an improvement, this definition still does not reflect well what should be understood as 'pro-poor growth' and falls short of providing straightforward answers to various plausible combinations of growth, poverty reduction, and inequality changes. Amartya Sen (1979, 1979c, 1979a, 1979b) in his Tanner Lecture at Standford on Human Development put forth 'capability approach' which is defined by its choice of focus upon the moral significance of individuals' capability of achieving the lives they have reason to value. This distinguishes it from more established approaches to ethical evaluation, such as utilitarianism or resource-ism, which focuses exclusively on subjective well-being or the availability of means to the good life, respectively. A person's capability to live a good life is defined in terms of the set of valuable 'beings and doings' like being in good health or having loving relationships with others to which they have real access. Functioning and capability were core ideas along with discourse of good life and basic conditions needed for good life. Nanak C. Kakwani (1980) in working paper published by World Bank has 'derived a poverty measure similar to that of Sen by an alternative approach of transfer of income from rich to poor so that the income of every poor is brought to the poverty line'.

Literature Gap

Existing literature at some extents examining the goals in particular region in especially homogenous in nature. This book attempt to fill the gap by examining heterogeneity in federal countries like India and South Africa. Further, the book explores the policies adopted at national and subnational level to achieve the dream of human development in paradigm of Sustainable Development Goals. This book will focus on the intervention made by the respective governments of India and South Africa in order to eradicate poverty, ensure availability of nutritious food for all, ensure quality education in order to enhance skill development and get better job opportunities, strengthening of health infrastructure regarding overall wellbeing, promote gender equality to ensure full participation of women in economic growth and to make people adapt to climate change.

The quantitative analysis will be based on household and individual information collected through surveys. For qualitative analysis, individual depth interviews and participant observation techniques will be adopted to obtain the information regarding SDGs target.

Relevance of the Study

SDGs are important milestone for progress and socio-economic development of countries with the pledge to "*leave no one behind by 2030*'. Before the advent of SDGs, MDGs led the socioeconomic progress of the nations. However, MDGs did not take in to the account the sustainable aspect of development. The aspects like climate change, quality education, overall wellbeing etc. were not adequately covered in the MDGs. At the same time, MDGs primarily covered developing countries while SDGs covers both developing and develop economies. SDGs are built on the success of MDGs. The goals of SDGs are meant for higher as well as lower income countries. The first five goals of SDGs are based on MDG1. Certain SDG goals correspond to the decent living conditions of the people. Some SDGs goals deal with climate change. The goals of SDGs are indivisible and balance the economic, social and environmental dimensions.

This study investigates the progress made by two emerging economies—India and South Africa–in realizing human development within the framework of the SDGs. In this effort, I traverse the stages in public policy from planning to implementation and from reach to range in the quest of attaining human development. My research includes cross-country qualitative as well as quantitative studies and in-depth analyses of India and South Africa.

India and South Africa are countries with unique federal structures that uphold the unbalanced states together. This study argues that uneven development challenges need asymmetric strategies, contrary to the axiom that SDGs are indivisible. Further advancement of India and South Africa depends on transforming their quantity (large youth population) into quality (human capital). Both countries are also striving to create platforms to unleash human potential, which is critically examined. The finding suggests that with improvements in health, education, and income, a rise in purchasing power can be anticipated on the demand side. On the supply side, public spending on education and health is indispensable for the enhancement of human capabilities. Both governments face the unique challenge of maintaining a balance between the uneven states/provinces while ensuring health, education, food, shelter, and decent jobs through targeted pro-poor policies.

The developing countries like India and South Africa have made considerable progress in achieving the goals of MDGs. However, there is a scarcity of researches that address the progression of India and South Africa towards the achievement of SDGs. This research attempts to make a contribution in this relatively neglected area of study.

This book is limited to India and South Africa and focuses on intervention made by the respective governments in order to eradicate poverty, ensure availability of nutritious food for all, ensure quality education in order to enhance skill development and get better job opportunities, strengthening health infrastructure for overall well-being, promote gender equality to ensure full participation of women in economic growth and to make people adapt to climate change. It aims to provide critical insights for better targeting of these programs.

Conclusion

World has travel long conceptualize the universal goals for human development. Starting from growth model to development and finally the sustainable development. All efforts aim to realize the targets of human development proposed by Mahbub Ul Haq and Amrtya Sen. Further the parameters of measuring poverty and deprivation refined by Sabina Alkire and Foster in concept of Multi-Dimensional Poverty Index from Human Development Index.

The first universal goals, i.e., Sustainable Development Goals were conceived at the United Nation conference on sustainable development in Brazil in 2012. These goals are based on the success of the Millennium Development Goals and to finish the unfinished targets. These goals are meant for all the countries, either higher income or lower income. First five goals of SDGs (No Poverty, Zero Hunger, Good Health and Well Beings, Quality Education and gender equality) are based on MDG1 (Eradicate extreme poverty and hunger). SDG8, SDG9 and SDG10 correspond to improving the living condition of the people by means of decent and new innovative ideas and sustainable economic growth while SDG 6, SDG 7 SDG11, SDG12, SDG13, SDG 14, SDG15 SDG 16 ADG 17 deals with climate change. These goals are achieved by focusing on each and every gals separately and balance the economic, social and environmental dimensions. These goals can be realized only by fostering gender equality, and not otherwise.

Sector specific assessment of progress toward MDGs identifies health problems, lack of quality education and income inequality as the major challenges towards the achievement of MDGs. Following chapters will shed light on these issues in detail. Chapter I focus on the necessity of sustainable development required in realizing human development in the true sense. This chapter highlights very specific nature of many goals, reflecting their diverse, independent origins, and the considerable gaps in coverage that could arise across their implementation. Chap. 2 examine the progress of India from MDGs to SDGs. India has registered considerable achievement in the targets of MDGs and is committed to achieve the goals of SDGs by 2030. This chapter identifies the gaps in the progress of respective goal and looks at targeted programs in order to achieve the SDGs. Chapter 3 analyzes the progress of South Africa from MDGs to SDGs. South Africa has scored well in reducing poverty eradication and eliminating hunger. The target of universal enrolment ratio has also been achieved. Yet the state of inequality reflects slanginess. This chapter discusses the impact of pro-poor programs in reducing poverty to zero% in order to bring equality. Chapter 4 discusses the parameters of SDGs in India and South Africa. This chapter focuses deliberately on problems with the SDGs to identify better future approaches and suggest a set of principles that might guide future goal development in South Africa and India. Chapter 5 provides a conclusion and explains the government policies focusing on increasing economic growth as this by itself reduces the cost of achieving the SDGs. It addresses the issues such as the quality-of-service delivery by government and improved monitoring of the SDGs achievements. It identifies challenges that are common to different goals in the achievement of SDGs.

References

Agarwal, N., Gneiting, U., & Mhlanga, R. (2017). *RAISING THE BAR: Rethinking the role of business in the Sustainable Development Goals*. Cowley, Oxford: OXFAM.

Akçay, S., Yiğit, R. R., & Bay, F. (2024). Experiences of child marriage among women and men in Turkey. *Journal of Child and Family Studies*.https://doi.org/10.1007/s10826-024-02791-5

Alkire, S., & Santos, M. E. (2010). *Acute multidimensional poverty: A new index for developing countries. OPHI Working Paper Series No. 38.*. Oxford University, Department of International Development, Oxford Poverty and Human Development.

Alkirea, S., & JamesFoster. (2011). Counting and multidimensional poverty measurement. *Journal of Public Economics*, 476–487.

Anand, S., & Sen, A. (1994). *Sustaible human development: Concept and priorities*. United Nations.

Andrews, M. A., Areekal, B., Rajesh, K., Krishnan, J., Suryakala, R., Krishnan, B., … Santhosh, P. (2020). First confirmed case of COVID-19 infection in India: A case report. *Indian Journal of Med Research*, 490–492.

Bandola-Gill, J., Grek, S., & Tichoner, M. (2022). *Governing the sustainable development goals. Quantification in global public policy (Sustainable Development Goals Series)* .. Springer Nature.

Banerjee, A. V., & Duflo, E. (2005). Growth Theory through the Lens of Development Economics. In P. Aghion, & S. N. Durlauf, *Handbook of Economic Growth*. Volume 1, Part A (pp. 473–552). Amsterdam: North-Holland.

Banerjee, A., & Duflo, E. (2012). Poor Economics: *Poor Economics, a radical rethinking of the way to fight global poverty*. New Delhi: Penguin Random House India.

Bauer, P. (1971). *Dissent on development: Studies and debates in development economics*. Weidenfeld and Nicolson.

Bourguignon, F., & Charavarty, S. R. (2003). The measurement of multidimensional poverty. *Journal of Economic Inequality*, 25–49.

Bradshaw, C. J., Ehrlich, P. R., Beattie, A., Ceballo, G., Crist, E., Diamond, J., … Wa, M. (2021). Underestimating the Challenges of Avoiding a Ghastly Future. *Frontier*, 1–10.

Bue, M. L., & Klasen, S. (2013). Identifying synergies and complementarities between MDGs: Results from cluster analysis. *Socila Indicators Research, 113*, 647–670. https://doi.org/10.1007/s11205-013-0294-y

Callaway, E. (2020). Time to use the p-word? Coronavirus enters dangerous new phase. *Nature*, https://doi.org/10.1038/d41586-020-00551-1.

Camilleri, M. A., & Camilleri, A. C. (2019). The sustainable development goal on quality education. In S. O. Idowu, R. Schmidpeter, & L. Zu (Eds.), *The Future of the UN Sustainable Development Goals: Business Perspectives for Global Development in 2030 CSR, Sustainability, Ethics & Governance* (pp. 261–277). Springer.

Castano-Isaza, J., & Diez, S. M. (2021). *Protecting oceans from climate change impacts*. Washington DC. Retrieve from: https://blogs.worldbank.org/climatechange/protecting-oceans-climate-change-impacts: The World Bank.

Costanza, R., Daly, L., Fioramonti, L., Giovannini, E., Kubiszewski, I., Mortensen, L. F., … Wilkinson, R. (2016). Modelling and measuring sustainable wellbeing in connection with the UN Sustainable Development Goals. *Ecological Economics, 130*, 350–355.https://doi.org/10.1016/j.ecolecon.2016.07.009

Dasgupta, A. K. (1954). Keynesian Economics and Underdeveloped Countries. *EPW*, 101–106.

Fan, X., & Sturman, M. (2019). Has higher education solved the problem? Examining the gender wage gap of recent college graduates entering the workplace. *Compensation & Benefits Review.*, *51*(1), 5–12. https://doi.org/10.1177/0886368719856268

Drèze, J., & Sen, A. (2013). *An Uncertain Glory: India and its Contradictions*. Princeton: Princeton University Press.

FAO. (2010). *Newest information and knowledge about the world's forests and forestry: Key Findings*. Food and Agriculture Organozations.

FAO. (2022). *The State of Food Security and Nutrition in the World 2022*. Rome: Food and Agriculture Organisation.

FAO, WTO, World Bank. (2023). *Rising global food insecuity: Assessing policy response.*. FAO, WTO, World Bank.

Filho, W. L., Wall, T., Salvia, A. L., Dinis, M. A., & Mifsud, M. (2023). The central role of climate action in achieving the United Nations' Sustainable Development Goals. *Scientific Reports volume 13, Article number: 20582*, https://doi.org/10.1038/s41598-023-47746-w.

Francesco Fuso Nerini, J. T., Spataru, C., Broto, V. C., Anandarajah, G., Milligan, B., & Mulugetta, Y. (2018). Mapping synergies and trade-offs between energy and the Sustainable Development Goals. *Nature Energy, 3,* 10–15.https://doi.org/10.1038/s41560-017-0036-5

Frieden, J. (2024). A Place for Politics. *Finance and Development*, pp. https://www.imf.org/en/Publications/fandd/issues/2024/03/A-Place-for-Politics-Jeffry-Frieden?utm_medium=email&utm_source=govdelivery.

Gasper, D. (2000). Development as freedom: Taking economics beyond commodities— The cautious boldness of Amartya Sen. *Journal of International Development*, *12*(7), 989–1001.

Gasper, D., Shah, A., & Tankha, S. (2019). The framing of sustainable consumption and production in SDG 12. *Global Policy, 10*(Suppl. 1), 83–95.

Gautam, S., & Hens, L. (2020). SARS-CoV-2 pandemic in India: What might we expect? *Environment, Development and Sustainability*, 3867–3869.
Giannetti, B., Agostinho, F., Almeida, C., & Huisingh, D. (2015). A review of limitations of GDP and alternative indices to monitor human wellbeing and to manage eco-system functionality. *Journal of Cleaner Production, 87*, 11–25. https://doi.org/10.1016/j.jclepro.2014.10.051
Grainger-Brown, J., & Malekpour, S. (2019). Implementing the sustainable development goals: A review of strategic tools and frameworks available to organisations. *Sustainability*, 1–18.
Gundry, S., Wright, J., & Conroy, R. (2004). A systematic review of the health outcomes related to household water quality in developing countries. *Journal of Water and Health*, 1–13.
Hansen, M., & Quintero, D. (2023). Public schools heavily rely on women's labor. Why do they pay female teachers less? *Brookings*, pp. https://www.brookings.edu/articles/gender-wage-gaps-policy-brief/.
Hill, P. S., Buse, K., Brolan, C. E., & Ooms, G. (2014). How can health remain central post-2015 in a sustainable development paradigm? *Global Health, 10*(18), 2–5. https://doi.org/10.1186/1744-8603-10-18
Huang, Y. (2021). Technology innovation and sustainability: Challenges and research needs. *Clean Technologies and Environmental Policy., 23*, 1663–1664. https://doi.org/10.1007/s10098-021-02152-6
IEA, IRENA, UNSD, World Bank & WHO. (2022). *Tracking SDG 7: The energy progress report.* World Bank.
ILO. (2022). *The gender gap in employment: What's holding women back?*. https://www.ilo.org/infostories/en-GB/Stories/Employment/barriers-women#intro. International Labour Organization.
ILO. (2023). *World employment and social outlook: Trends 2023.* International Labor Organizations.
IMF. (2024). *Real GDP growth: Annual Percent Change.* International Monetary Fund.
(1980). *Income Inequality and Poverty: Method of Estimation and Policy Appication.*. The World Bank.
Indian Express. (2021). Covid-19 India timeline: Looking back at pandemic-induced lockdown and how the country is coping with the crisis. *Indian Express*, pp. Available at: https://indianexpress.com/article/india/covid-19-india-timeline-looking-back-at-pandemic-induced-lockdown-7241583/.
IPU. (2024). *Monthly ranking of women in national parliaments.*. International Parliamentary Union.
Islam, S. N. (2020). *Rivers and sustainable development: Alternative approach and their implication.* Oxford University Press.
Jin, Y., Yang, H., Ji, W., Wu, W., Chen, S., Zhang, W., & Duan, G. (2020). Virology, epidemiology, pathogenesis, and control of COVID-19. *Viruses*, 1–17.
Jong, W. D., Liu, J., & Long, H. (2021). Wil de Jong,corresponding author1 Jinlong Liu,1 and Hexing Long. *Ambio, 50(12)*, 2224–2237.
Keitsch, M. M., & Vermeulen, W. J. (2020). *Transdisciplinarity for sustainability: Aligning diverse practices.* Routledge.
Khan, I., Haleem, A., & Javaid, M. (2020). Analysing COVID-19 pandemic through cases, deaths, and recoveries. *Journal Oral Biol Craniofac Res*, 450–469.
Kroll, C., Warchold, A., & Pradhan, P. (2019). Sustainable Development Goals (SDGs): Are we successful in turning trade-offs into synergies? *Palgrave Commun, 5*, 140. https://doi.org/10.1057/s41599-019-0335-5
Lagarde, M., Haines, A., & Palmer, N. (2007). Conditional cash transfers for improving uptake of health interventions in low- and middle-income countries: A systematic review. *JAMA.*
Lau, H., Khosrawipour, V., Kocbach, P., Mikolajczyk, A., Schubert, J., Bania, J., & Khosrawipour, T. (2020). The positive impact of lockdown in Wuhan on containing the COVID-19 outbreak in China. *Journal of Travel Medicine*, 1–25.
Leggett, J. A., & Carter, N. T. (2012). *Rio+20: The united nations conference on sustainable development, June 2012* . Congressional Research Service.
Lomazzi, M., Borisch, B., & Lasser, U. (2014). The Millennium Development Goals: experiences, achievements and what's next. *Global Health Action*, 1–9.

Long, W. J. (2021). The Aśokan Empire. In W. J. Long, *A Buddhist Approach to International Relations* (pp. 51–70). Springer Nature.

Lu, Y., Nakicenovic, N., Visbeck, M., & Stevance, A.-S. (2015). Policy: Five priorities for the UN sustainable development goals. *Nature, 520*, 432–433. https://doi.org/10.1038/520432a

Lusseau, D., & Mancini, F. (2019). Income-based variation in sustainable development goal interaction networks. *Nature Sustainability, Nature,* 2(3), 242–247. https://doi.org/10.1038/s41893-019-0231-4

Mandelbaum, K. (1945). *The industrialisation of backward areas.*. Basil Blackwell.

Markus Kaltenborn, M. K. (2020). *Sustainable development goals and human rights.*. Springer Nature. Retrieve from: https://doi.org/10.1007/978-3-030-30469-0.pdf.

Martíneza, J., Mboupb, G., Sliuzasa, R., & Stein, A. (2008). Trends in urban and slum indicators across developing world cities, 1990–2003. *Habitat International*, 86–108.

Mathy, S., & Blanchard, O. (2016). Proposal for a poverty-adaptation-mitigation window within the Green Climate Fund. *Climate Policy, 16*(6), 752–767. https://doi.org/10.1080/14693062.2015.1050348.

McMichael, A. J., S Friel, F., A Nyong, D., & Corvalan, C. (2008). Global environmental change and health: Impacts, inequalities, and the health sector. *BMJ*, 191–194.

Mekonnen, M. M., & Hoekstra, A. Y. (2016). Four billion people facing severe water scarcity. *Science Advances, 2*(2), https://doi.org/10.1126/sciadv.150032.

Min Cao, M. C., Chang, L., Chen, Y., Sun, Z., Xu, Z., Zhu, R., … Lü, G. (2023). 2023. *Humanities and Social Sciences Communications, 10*, Article number: 450, 1–9.https://doi.org/10.1057/s41599-023-01952-z

Moon, B. K. (2015). Foreword. In UNDP, *The Millennium Development Goals Report 2015* (p. 3). UNDP.

MOSPI. (2017). *Millennium development goals- final country report India.*. Ministry of Statistics and Implementation. Government of India.

Nilsson, M., Griggs, D., & Visbeck, M. (2016). Policy: Map the interactions between Sustainable Development Goals. *Nature, 534*, 320–322. https://www.nature.com/articles/534320a.

Nordhaus, W. D., & Tobin, J. (1973). Is growth obsolete. In M. Moss, *The Measurement of Economic and Social Performance* (pp. 509–564). NBER. Retrieved from: http://www.nber.org/books/moss73-1.

Nunes, A. R., Lee, K., & O'Riordan, T. (2016). The importance of an integrating framework for achieving the Sustainable Development Goals: the example of health and well-being. *BMJ Global Health*, 1–12.https://doi.org/10.1136/bmjgh-2016000068

Nurkse, R. (1961). *Problems of capital formation in underdeveloped countries.* Oxford University Press.

Nussbaum, M. C. (2013). *Creating Capabilities: The Human Development Approach.* Belknap Press

OECD. (2017). *Policy Coherence for Sustainable Development 2017. Eradicating Poverty and Promoting Prosperity.* Paris: OECD

Pine, T. (1792). *Rights of man. Part the Second.*. Retrieve from: https://pdcrodas.webs.ull.es/anglo/PaineRightsOfMan.pdf.

Prachitha, J., Dhume, A., & Subramanian, S. (2019). India in Pursuit of millennium development goals: Were the targets really feasible? *Journal of Developing Societies*, 105–133.

Pradhan, P., Costa, L., Rybski, D., Lucht, W., & Kropp, J. P. (2017). *Advancing earth and space science.*https://doi.org/10.1002/2017EF000632

Rangarajan, L. (1987). *Kautilya. The Arthshastra.* Penguin Books.

Ravallion, M. (2004). *Pro-poor growth: A primer. World Bank Policy Research Working Paper 3242,*. Washington DC: The World Bank.

Reddy, S. (2018). *Plastic Pollution Affects Sea Life Throughout the Ocean.* PEW.

Rehfuess, E., Mehta, S., & Prüss-Üstün, A. (2006). assessing household solid fuel use: Multiple implications for the millennium development goals. *Environ Health Perspect*, 373–378.

Remuzzi, A., & Remuzzi, G. (2020). COVID-19 and Italy: what next? *Lancet*, 1225–1228.

Renaud, F. G., Zhou, X., Bosher, L., Barrett, B., & Huang, S. (2022). Synergies and trade-offs between sustainable development goals and targets: Innovative approaches and new perspectives. *Sustainability Science., 17*, 1317–1322.

Rickels, W., Dovern, J., Hoffmann, J., Quaas, M. F., Schmidt, J. O., & Visbeck, M. (2016). Indicators for monitoring sustainable development goals: An application to oceanic development in the European Union. *Earth's Future., 4*(5), 252–267. https://doi.org/10.1002/2016EF000353

Rolland, J. S. (2020). COVID-19 Pandemic: Applying a multisystemic lens. *Family Process*, 922–936.

Rosenstein-Rodan, P. N. (1943). Problems of industrialisation of Eastern and South-Eastern Europe. *The Economic Journal*, 202–211.

Sachs, J. D. (2012). From millennium development goals to sustainable development goals. *The Lancet, 379*(9832), 2206–2211. https://doi.org/10.1016/S0140-6736(12)60685-0

Sassler, S., & Meyerhofer, P. (2023). Factors shaping the gender wage gap among college-educated computer science workers. *PLoS One, 18(10), e0293300*, https://doi.org/10.1371/journal.pone.0293300. PMID: 37903156; PMCID: PMC10615266.

Schoch, M., & Lanker, C. (2020). The number of poor people continues to rise in Sub-Saharan Africa, despite a slow decline in the poverty rate. *Data Blog*, pp. https://blogs.worldbank.org/opendata/number-poor-people-continues-rise-sub-saharan-africa-despite-slow-decline-poverty-rate#:~:text=This%20page%20in%3A&text=This%20translates%20to%2040%25%20of,of%20poor%20continues%20to%20rise.

Seferian, S. M. (2021). *Sustainable Minimalism: Embrace Zero Waste, Build Sustainability Habits That Last, and Become a Minimalist without Sacrificing the Planet.* Mango.

Sen, A. (1979a). *Equality of what? The tanner lecture on human develoomet.* Stanford University.

Sen, A. (1979b). Issues in the measurement of poverty. *The Scandinavian Journal of Economics*, 285–307.

Sen, A. (1993). Capability and well-being. In M. Nussbaum & A. Sen (Eds.), *The Quality of Life* (pp. 30–53). Oxford University Press.

Sen, A. (2010). A decade of human development. *Journal of Human Development*, 17–23.

Shulla, K., & Kőszeghy, L. (2021). *SDG BOOKLET: Housing ensures sustainable development.* Bratislava: Habitat for Humanity.

Smith, P., & Olesen, J. E. (2010). Synergies between the mitigation of, and adaptation to, climate change in agriculture. *The Journal of Agricultural Science*, 543–552.https://doi.org/10.1017/S0021859610000341

Solow, R. (1992). An almost practical step toward sustainability. Extract from lecture given by Professor Solow in 1992 to a meeting at Resources for the Future, Washing DC. *Ekistics*, 15–20.

Spiggle, T. (2021). The Gender Pay Gap: Why It's Still Here. *Forbes*, pp. https://www.forbes.com/sites/tomspiggle/2021/05/25/the-gender-pay-gap-why-its-still-here/?sh=3f0d9def7baf.

STATS SA. (2016). *Millennium Development Goals: Country Report 2015.*. Statistics South Africa.

Stiglitz, J. (2009). GDP Fetishism. *Project Syndicate*, pp. https://www.project-syndicate.org/commentary/gdp-fetishism.

Stiglitz, J. E., Sen, A., & Fitoussi, J.-P. (2009). *Report by the commission on the measurement of economic performance and social progress.* . OFCE—Centre de recherche en économie de Sciences Po. https://sciencespo.hal.science/hal-01069384/document.

Subramanee, S. D., Agho, K., Lakshmi, J., Huda, M. N., Joshi, R., & Akombi-Inyang, B. (2022). Child Marriage in South Asia: A systematic review. *International Journal of Environmental Research Public Health, 19*(22), 15138. https://doi.org/10.3390/ijerph192215138.

Swingler, G., Hendricks, M., Hall, D., Hall, S., Sanders, D., McKerrow, N., … Reid, S. (2012). Issues in Child Health: Can a new paediatric sub-specialty improve child health in South Africa? *South African Medical Journal*, 738–739.

UIS. (2019). *New methodology Shows that 258 million children, adolescents and youth are out of school.* UNESCO Institute for Statistics.

UN. (1987). *United nations brundtland commission report.*. United Nations. http://www.un-documents.net/our-common-future.pdf.

Women, U. N. (2015). *Progress of the world's Women 2015–2016: Transforming economies, realizing rights*. UN Women.

UN Women. (2022). Press release: Achieving full gender equality is still centuries away, warns the United Nations in a new report. *PRESS RELEASE*, pp. https://www.unwomen.org/en/news-stories/press-release/2022/09/press-release-achieving-full-gender-equality-is-still-centuries-away-warns-the-united-nations-in-a-new-report.

UNDP. (1990). *Human development report*. United Nations DEvelopment Programme.

UNDP. (2010). *Human development report 2010*. United Nations Development Programme.

UNEP. (2021). *Food Waste Index Report 2021.*. United Nations Environment Programme.

UNESCO. (2015). *Incheon declaration and SDG4—education 2030 framework for action*. United Nations Educational, Scientific and Cultural Organization.

UNESCO. (2017). *Education for sustainable development goals: Learning objective 2030.*. UNESCO.

UNESCO. (2023). One year into COVID-19 education disruption: Where do we stand? *News*, pp. https://www.unesco.org/en/articles/one-year-covid-19-education-disruption-where-do-we-stand.

UNHCR. (2022). *Refugee data finder*. Geneva. Retrieve from: https://www.unhcr.org/refugee-statistics/: UNHCR. The Refugee Agency.

UNICEF & WHO. (2023). *Progress on household drinking water, sanitation and hygiene 2000–2022: special focus on gender.*. United Nations Children's Fund (UNICEF) and World Health Organization (WHO).

UNICEF. (2014). WHO/UNICEF highlight need to further reduce gaps in access to improved drinking water and sanitation. UNICEF.

UNICEF. (2021a). 10 million additional girls at risk of child marriage due to COVID-19—UNICEF. *Press Release*, pp. https://www.unicef.org/eap/press-releases/10-million-additional-girls-risk-child-marriage-due-covid-19-unicef.

UNICEF. (2021b). Child labour rises to 160 million—first increase in two decades. *Press Release*, pp. https://www.unicef.org/press-releases/child-labour-rises-160-million-first-increase-two-decades.

United Nations. (2000). SDG 15. Life on Land. In U. Nations, *Sustaiable Development Goals Report 2022* (pp. 56–57). United Nations.

United Nations. (2015). *The Millennium development goals report: 2015.*. United Nations.

United Nations. (2021). *Progress towards the Sustainable Development Goals. Report of the Secretary-General.*. United Nations.

United Nations. (2022a). *Ensure inclusive and equitable quality education and promote lifelong learning opportunities for all.*. Retrived from: https://unstats.un.org/sdgs/report/2020/Goal-04/: United Nations.

United Nations. (2022b). SDG3 helath and wellbeings. In U. Nations (ed.), *The Sustainable Development Goals Report 2022* (pp. 30–34). United Nations.

United Nations. (2022c). *UNSTAT: SDG 9*. New York: Department of Economic and Social Affairs. United Nations.

Walker, J., & Pekmezovic, A. (2019). *Sustainable development goals: Harnessing business to achieve the SDGs through finance*. Wiley.

Walker, P. J. (2018, March). Peter J. Walker profiles angus deaton, who connects the dots between theory, policy, and people's lives. *Finance and Development*, pp. https://www.imf.org/en/Publications/fandd/issues/2018/03/people#:~:text=He%20also%20analyzed%20how%20consumption,income%20should%20lead%20to%20larger.

Wan, G., & Wang, C. (2019). Poverty and inequality. In D. N. (ed), *Asian Transformations: An Inquiry into the Development of Nations* (pp. 240–266. https://doi.org/10.1093/oso/9780198844938.003.0010). Oxford Academic.

WHO & World Bank. (2023). *Tracking universal health coverage 2023 global monitoring report*. WHO & World Bank.

WHO. (2020a). Novel Coronavirus (2019-nCoV) SITUATION REPORT—2. *World Health Organization*, pp. Available at 20200122-sitrep-2–2019-ncov.
WHO. (2020b). Statement on the second meeting of the International Health Regulations (2005) Emergency Committee regarding the outbreak of novel coronavirus (2019-nCoV). *World Health Organization*, pp. Availble at: https://www.who.int/news/item/30-01-2020-statement-on-the-second-meeting-of-the-international-health-regulations-(2005)-emergency-committee-regarding-the-outbreak-of-novel-coronavirus-(2019-ncov).
WHO. (2020c). *WHO Director-General's opening remarks at the media briefing on COVID-19–11 March 2020.* World Health Organization.
WHO. (2021). Violence against women. *Fact Sheets*, pp. https://www.who.int/news-room/fact-sheets/detail/violence-against-women#:~:text=Scope%20of%20the%20problem&text=Over%20a%20quarter%20of%20women,lifetime%20(since%20age%2015).
WHO. (2022). *Drinking-water.*. World Health Organization.
WHO. (2023). *World health statistics 2023.*. World Health Organizations.
Wollstonecraft, M. (1792). *Vindication of the rights of woman.*
World Bank. (2020a). *The world bank's education response to COVID-19* . World Bank.
World Bank. (2022b). *Global economic prospects 2022.* World Bank.
World Bank. (2022c). *Poverty and shared prosperity 2022: Correcting course.* World Bank.
World Bank. (2022d). *What You need to know about food security and climate change.* World Bank.
Yonzan, N., Mahler, D. G., & Lanker, C. (2023). *Poverty is back to pre-COVID levels globally, but not for low-income countries.* World Bank. https://blogs.worldbank.org/opendata/poverty-back-pre-covid-levels-globally-not-low-income-countries.
Zhu, N., Zhang, D., Wang, W., Li, X., Yang, B., Song, J., … Tan, W. (2020). A Novel Coronavirus from Patients with Pneumonia in China, 2019. *The New England Journal of Medicine*, 727–733.

Chapter 2
Implementation of Sustainable Development Goals in India: Progress and Challenges

Introduction

The quest for sustainable development is rooted in Indian tradition. India represents a culture where the earth is regarded as mother, embedded deep in ancient Indian texts that the earth should be kept pure for she is our mother! And we are her children. India's great son Mahatma Gandhi said that there is enough for everybody's needs but not enough for everybody's greed. Gandhian worldview is still playing an important role in guiding policy formulation in India. At the UN Summit for the adaptation of SDGs, Prime Minister of India, Narendra Modi started his speech by quoting Mahatma Gandhi "one must care about the world one will not see" (NITI Aayog, 2017).

The Gandhian approach to human development was carried forward by his disciple Pt. Jawahar Lal Nehru, the first Prime Minister of India. When India breathed the first air of freedom, Pt. Nehru on the eve of independence in his Constituent assembly speech famously uttered, "*We take the pledge of dedication to the service of India and her people and to the still larger cause of humanity*" (Nehru, 1947). The challenge before the newly independent Indian nation was not merely to identify the goals for a new country but also to map the challenges in the path of humanity. For Nehru, freedom entailed a step forward to eradicating poverty and ignorance, reducing the burden of diseases, and equal opportunity without any discrimination. His idea of freedom encapsulated the paradigm of 'development'. Freedom is key to development. One of the finest scholarships of Amartya Sen, *Development as Freedom,* complements Nehru's idea of freedom wherein Sen advocated for the expansion of freedom, both as the primary end and core of development (Sen, 1999). Sen contended that the meaning of freedom for an individual is to pursue his/her actual ability to achieve the functioning necessary for a decent life (Sen, 1993). Sen held that the transformation of individuals' capability into functioning, to some extent, relies on public policies in advancing human development. The philosophy

U. K. Singh, *Regional Dimensions of Human Development in India and South Africa*, Advances in Geographical and Environmental Sciences,
https://doi.org/10.1007/978-981-97-5846-3_2

of Mahatma Gandhi, the idea of freedom of Nehru, and the approach of Amartya Sen find expression in Agenda 2030, a set of goals, to make a new humane world.

Sustainable development goals (SDGs) are a step forward to redeem the idea of freedom, envisaged by great leaders of Indian independence. India has learned a lot from the experiences of MDGs, the goals are indicators to measure time-bound objectives concerning poverty, hunger, disease, schooling, gender inequality, and deteriorating environment. Emerging environmental challenges raise widespread understanding to prioritize the environmental objectives in coherence with the poverty reduction objectives. It is an attempt to free the world from hunger, ensure better health and education, and achieve gender equality. The SDGs are also an immediate call to make a more sustainable world by 2030 (UN, 2015). Unlike MDGs, the SDGs apply to all countries (Glaser, 2012; Singh, 2022a, 2022b). Its 17 goals, 169 targets, and 231 indicators expand over the social, economic, and cut across the social, economic, and environmental dimensions of suitability. Advancement in science and technology has brought the world closer and the spirit of partnership between member countries remains crucial for achieving SDG goals and building a new world by 2030.

The Indian economy has shown astonishing progress in achieving robust economic growth. The success of India is marked by achieving the important transition from a food importer to a food exporter and becoming a leading force in space and satellite technology, among other praiseworthy advances. Throughout this momentous journey, India has maintained its vibrant and secular democratic beauty, even though rising inequality, lack of decent jobs and poverty remain subjects of introspection. The journey of India stands as a testament to a country marching on the path of progress by maintaining its core idea of '*vasudhaiv kutumbkam*' (the world is one family), an idea that needs to be revived and nurtured through the Sustainable Development Goals -2030. As a signatory of the global goals under Agenda -2030, India has pledged to finish the unfinished goals of the Millennium Development Goals manifested in the form of Sustainable Development Goals (SDGs).

Implementation of Sustainable Development Goals in India

The world has travelled halfway towards SDGs and it has become evident that the success of SDGs to some extent depends on the progress that India makes. One-sixth of humanity lives in India, comprising 18% of the global youth population. India is experiencing the largest ever population of adolescents and youth, one in every five persons is an adolescent under the age of 10–19 years and nearly 370 million people, one in every three persons is a young person in the age group 10–24 years. India is experiencing a 'youth bulge', a demographic window opportunity that will continue till 2025. Further advancement of India depends on transforming its quantity (large youth population) into quality (human capital). It is understood that the relative success or failure of SDGs also, to some extent, relies on India's performance, the most populated country where 14.96% of the population is still living in abject multi-dimensional poverty (NITI Aayog, 2023a, 2023b). The triumph of SDGs depends on the localization of goals, by identifying national challenges like hunger, poverty, sustainable growth, gender equality, quality education, and environment along with other several issues at national and sub-national levels. In this context, India's commitment to the implementation of SDGs and converging the targets with local goals assume critical significance.

India's final report on MDGs shows that the country has marked significant progress in halving poverty, and improving public health, education, and environment-related goals in MDGs. (MoSPI, 2017) However, targets of hunger, maternal health, and sanitation show considerable results and seek extra attention (Verma, 2017; Prachitha et al., 2019). Although women's participation in decision-making has achieved some success, the target of fifty percent at the national and sub-subnational levels remains a challenge (Singh, 2022a, 2022b). To complete the unfinished task of MDGs, India has adopted the SDGs envisaging a robust development by 2030. The prime focus of SDGs is to shift the economy on a sustainable path.

India is a diverse country where priorities of development differ from state to state. Every state is unique in its strengths and challenges. The task of recognizing the potential and weakness at the sub national is a difficult task. The National Institute of Transforming India (NITI) Ayog is a 'think tank' set up by the Government of India to identify the challenges of development at the national and sub national level in India. This agency is mandated to suggest the government at all levels of governance on the issues of development with targeted programme and policies and to take India on the path of sustainable development.

NITI Aayog is tasked to facilitate the collaboration between various stake holders like union and state governments, the designated agency to monitor the progress and coordination for achieving SDGs. Another central agency involved in SDGs is the Ministry of Statistics and Programme Implementation (MoSPI). This ministry is assigned to frame indicators to release National Indicators Framework (NIF) by engaging ministries at the national and sub-national levels along with international stakeholders. Both agencies, NITI Aayog and MoSPI work to identify priority indicators at the national and state levels and converge them to the 169 targets of SDGs (NITI Aayog, 2018a, 2018b). SDG 12, SDG13, SDG14, and SDG17 are excluded from the tabulation of the index due to the unavailability of a comparable data set. The first SDGs Index was released in 2018.

Within three years, the efforts of the agencies are reflected in the third edition of the SDGs Index of India. The number of indicators increased to 115 covering 70 targets across the 17 goals. The expansion of monitoring indicators shows the country's concern for accelerating goals (Kant, 2021).

The second SDG India Index (2021) report shows that at the national level India is on the path to achieving the SDGs within the stipulated period. Analyzing the two consecutive SDG India report (See Fig. 2.1) the score of India at the national level increased from 60 in 2019–20 to 66 in 2020–21. The improvement in scores shows the positive trends towards the attainment of SDGs. Trends are improving and satisfactory for the goals. The score for SDG3, SDG6, SDG7, SDG10, SDG11,

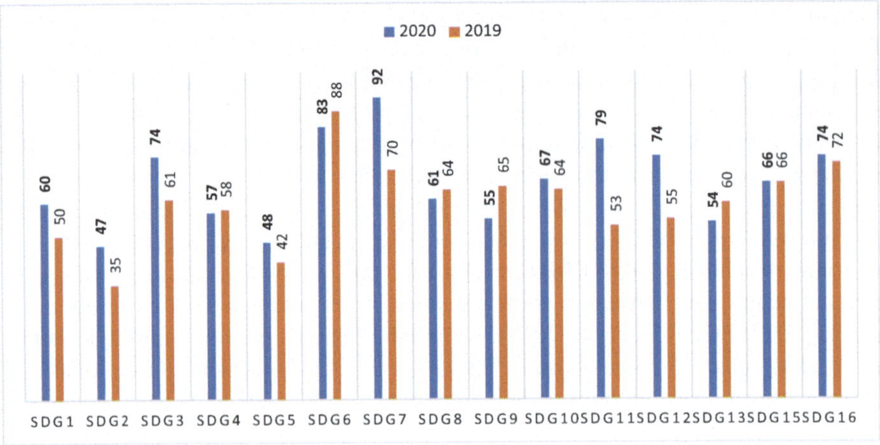

Fig. 2.1 SDG Index India Report; 2020–21 and 2018–19. *Source* NITI Aayog (2021a, 2021b)

SDG12, SDG15, SDG16 are in the front runner category with score 65–99, and scores for SDG1, SDG4, SDG8, SDG9 and SDG13 are moving vertically, their scores fall under performer category with score 50–64, while SDG2, SDG3, and SDG5 need immediate attention since their scores are below 50 which places them in aspirant category. The performance of individual indicators relating to specific goals is given in Table 2.1. Scores for SDG 14 (Life below sea) and SDG 17 (Partnership for goals) are not available at the national level since SDG 14 is related only to coastal states and SDG 17 is qualitative in nature.

Progress and targets of priority indicators are given in Table 2.1. The evidence suggest that India is on the path of attaining SDGs by 2030. But, COVID-19 Pandemic has disrupted the path of achieving these goals in more ways than imagined. The global economic slowdown in an interconnected world has hampered the progress of developing countries like India with a large population.

Assessment of Sustainable Development Goals in India

This chapter will discuss the progress achieved in the first three years of SDG's implementation. It explores the impact of the pandemic on SDGs in India. It also deciphers the gap between the stimulus package and needs on the ground. Further, its asses the government intervention in the advancement of SDGs. And in last, it attempts to propose a way forward. This chapter identifies no poverty, zero hunger, health, quality education, gender equality, water and sanitation, inequality, and decent jobs for further discussion.

Table 2.1 Selected indicators of undernutrition

Countries	Prevalence of undernourishment in the total population		Prevalence of wasting in children < 5 years	Prevalence of stunting in children < 5 years		Prevalence of Anaemia in women (15–49 years)		Prevalence of low birthweight	
	2004–06	2020–22	2022	2012	2022	2012	2019	2012	2020
India	21.4	16.6	18.7	41.6	31.7	53.2	53.0	29.5	27.4
China	7	<2.5	1.9	7.6	4.6	14.8	15.5	5.1	5
Bangladesh	13.7	11.2	9.8	39.2	26.4	35.7	36.7	24.3	23.0

Source FAO, IFAD, UNICEF, WFP & WHO (2023)

SDG1: No Poverty

The perception of poverty varies across geographies and across time. (Tripathi & Yenneti, 2020) However, a core element of measuring poverty remains the same across the countries, a poverty line, obtained by comparison of measurement of a household's income and cost of consumption required to meet the necessities to maintain a basic standard of living.

The massive economic growth of India, especially in post-reform era left an impressive mark on the global economy. It exerted a significant role in the rise of Asia in the global economy. Deepak Nayyar (2019) contended that India's growth rate had occupied second lead, after China among the Asian-14 group of countries. However, realizing high economic growth cannot be the prime objective of economic policy, but rather the redistribution of benefits from high growth rate must be the end goal of policy making. A high growth rate is unquestionably an important factor in poverty reduction but to make it sustainable it should be broad-based across sectors and (Ianchovichina, 2012). Fast and sustained poverty reduction needs inclusive growth that opens the door for people to contribute and benefit from the rapid economic growth and inclusive to all working hands.

Poverty is more than inadequacy of income; it is a multifaceted phenomenon. Thus, being poor, a person can face simultaneous challenges encompassing vulnerability and deprivation, such as hunger or malnutrition, poor health, low level of education, informal jobs, lack of clean water & sanitation, clean fuel and roof on head. Poverty is also about the lack of opportunity to achieve the overall wellbeing. The United Nation Development Program (UNDP) has deployed the Multi-Dimensional Poverty Index to measure the progress of countries against SDG 1 to end poverty in all its forms.

UNDP report (2019) highlighted that among the following 10 countries—Bangladesh, Cambodia, India, Niger, Pakistan, Vietnam, Democratic Republic of Congo, and Haiti—belonging to different income levels and comprising 2 billion population, the fastest absolute reduction in multidimensional poverty was visible in India, Cambodia, and Bangladesh, followed by Ethiopia and Haiti.

The national target of India regarding SDG1 is to reduce the share of poor to 10.95 by national poverty line and proportion of people living in abject of MPI is to 13.95 (NITI Aayog, 2021a, 2021b). The SDG Base line report demonstrate that the proportion of poor according to national poverty line is 21.92% in 2018 (NITI Aayog, 2018a, 2018b). The rural–urban divide is glaring in India. Nearly 21.2% of rural population are poverty ridden, compared to 5.5% in urban India. Rural India accounted for almost 90% of all poor people in India, the figure is 205 million out of 229 million based on the population data of 2020 (UNDP, 2022).

India has vast experience in poverty measurement. Earlier in India, poverty was estimated at the national and sub-national levels by the Planning Commission on a quinquennial basis. The last figure of poverty was released in 2013. As per the last Planning Commission report on poverty during 2011–2012, 21.9% of Indians were living below the national poverty line, Rs 816 per person per month in rural areas and Rs 1000 per person per month in urban areas. The number of poor in India further increased to 269.3 million (PIB, 2013). In 2022, United Nations literature noted that India is home to the world's largest poor population with 229 million people still living in poverty (UN, 2022). The Planning Commission also recognized the inter-state price differential, so the poverty line varies from state to state.

In 2021, Srinivas Kesineni, a Member of Parliament Lok Sabha, raised three questions in the Parliament regarding poverty in India, (i) details of poverty line, (ii) reason for delay in updating poverty line, and (iii) whether the government is considering new factors in deterring poverty line in India. In response, the Ministry of Rural Development stated that after the abolition of the Planning Commission, the Government relied on the Socio-Economic Caste Survey Census (2011) till the release of the first Multidimensional Poverty Index (MPI) by NITI Aayog, in November 2021 (Lok Sabha, 2021).

Recent MPI report statistics shows that 14.55% people are multidimensionally poor, 229 million people still living in poverty. The proportion of multidimensional poor peoples are starker in populace states like Bihar (33.76%), Uttar Pradesh (22.9%), Madhya Pradesh (20.63%) and (Rajasthan 15.31%) (NITI Aayog, 2023a, 2023b).

India's last poverty data shows that nearly 80% of the poor lives in rural areas (PIB, 2013). Mainstreaming of this large deprived population is the major concern of Indian leadership and policy makers. There are many ways to generate employment opportunities for poor, particularly minimum wage programme. The most effective way is a public employment programme with cash wages (Sen, 1998). In 2005, India enacted the world's largest public job guarantee programme, Mahatma Gandhi National Rural Employment Guarantee Act (MGNREGA). The programme was implemented in 2006 with the provision of a guaranteed 100 days job in a financial year to every rural household with members above 18 years of age who are willing to do unskilled annual work (PIB, 2021). The aim was to ease the burden of unemployment and hence poverty. MGNREGA is a demand-driven programme without any capacity limitation. The jobs include cleaning and digging wells, canals, and roads, cleaning weeds, tilling, ploughing, and so on. The annual outlay of the programme is nearly $10 billion per year, 2.5 billion days of employment in a financial year (Das & Mocan, 2012).

In the first five year of implementation of MGREGA, India witnessed a significant decline in the share of extreme poverty. The people's participation in the programme soared over the years. The monthly per capita expenditure on food among the beneficiary's household increased by 9.8% (Ravi, 2014). MGNREGA increased the capital expenditure by 10% (Deininger & Liu, 2013a, 2013b). For the first time, India saw a reduction in the number of poor during 2004–05, 2009/10, and 2011–12 which were 407 million, 355 million, and 270 million respectively (Balakrishnan, 2022). In 2023, the NITI Aayog report shows India's remarkable progress in lowering the poverty headcount ratio. India has successfully reduced the share of people in multidimensional poverty from 25.85% to 14.96% between 2015–16 and 2019–212,021 (NITI Aayog, 2023a, 2023b). In the last fifteen years the share of persons in multidimensional poverty fell from 55.1% to 16.4% during 2005–2006 and 2019–2021, with nearly 415 million people poor people uplifted from poverty (UNDP & OPHI, 2023). Notwithstanding the significant pre-pandemic progress, India is home to the world's largest poor population with 229 million people still living in poverty (UN, 2022).

Meanwhile, the participation in MGREGA reached 84.75% and 91.38% women are estimated to have benefitted from the *Pradhan Mantri Matru Vandan Yojna* (PMMVY). However, only 28.7% Indian household are covered by health insurance against the target of 100%. In an attempt to improve health insurance coverage, in 2018 the government of India launched *Pradhan Matri Ayushman Bharat Yajana*. Since its implementation it provided cashless treatment to 12.6 million beneficiaries (World Bank, 2022). This programme has the potential to improve insurance coverage in India.

SDG2: Zero Hunger and Undernutrition

Zero hunger is the second commitment of SDGs. SDG 2 aims to promote healthy and active life by eradicating hunger and making the world food secure. SDG2 mandates include supporting livelihood of small and marginalized farmers, encouraging the sustainable ways of agriculture without harming natural resources. Hunger can be explained in many ways. FAO defines hunger as an uncomfortable situation caused by lower dietary intake. Hunger is also linked to food insecurity. People may have food access to meet the required energy level but they are uncertain that it will last, or they are forced to reduce the quality and/or quality of diet they eat in order to get by. Hunger and undernutrition have long lasting impact on children under the age of 5 years.

Every one out of five children in the world are undernourished (Fontaine, et al., 2023). It is one of the major challenges to public health in world. It effects mainly infant and children. Wasting (low weight for height), stunting (low height for age) and being overweight are indicators used to measure the nutritional status. In spite of sustained high growth rate, undernutrition among children under five year of age is a major cause of child mortality in urban and rural areas (Murarkar et al., 2020). Around forty five percent deaths in children under five years of age is linked to undernutrition (WHO, 2021a, 2021b). An undernourished child is prone to infectious disease and it influence the overall development of child. In the long run childhood stunting leads to high risk for adult diseases like obesity, diabetes, cardiac and many more (UNICEF, WHO, & World Bank, 2023). The targets of SDG 2 and SDG 3 are committed to promote well beings to all. FAO report (2023) revealed that in 2022 there were nearly 148 million children under five years (22.3%) of age identified stunted growth, nearly 45 million (6.8%) were fell under wasted and 37 million (5.6%) who are overweight. Further, report observed that India is home to nearly 21.5 million (18.7%) children affected by wasting, highest in the world and 36.1 million (31.7%) children who are stunting children. Table 2.1 demonstrates that the share of undernourished people in India reduced from 21.4% to 16.6% during 2004–06 and 2020–22 in comparison to

China's remarkable achievement from 7% to below 2.5% during the same interval; although in Bangladesh proportion of undernourished fell from 13.7 to 11.2 the same period.

SDG3: Health & Wellbeing

Ensuring healthy life is vital for human being and the development of society. Michael Grossman explained this causality in a model whereby a person improves their stock of health by consuming healthcare, which in turn, returned in to a productive resource (Grossman, 1972). Good health is an indicator of a strong nation and better health sector is a prerequisite for a stable and prosperous economy (Bloom & Canning, 2003). This is because good health facilitates smooth functioning of economy and play a considerable role in the progress of socio-economic development. Studies have shown that the health sector play a constructive role in nation building (Huang et al., 2022). World Bank report (1993), '*Investing in Health*' identified public health as public good, which is complemented by Amartya Sen who included health as a key factor of social justice (Sen, 2002). Achieving healthy life and ensuring wellbeing is visualized in SDG3.

SDG3.1: Maternal Mortality Rate (MMR)

The healthy future of a country is conceived by every healthy mother, if a mother die, future also die with her. Bill & Melinda gates foundation report (2023) contends that maternal mortality has the potential to accelerate the incidence of infant mortality as nearly 67% babies died within a year of birth. Maternal mortality rate (MMR) indicates the degree of improvement in health and social and economic development of a country (Wilmoth et al., 2012). It is the measure of deaths due to complications during childbirth. SDG3.1 is committed to reduce the maternal mortality rate by 70 deaths/100,000 live births by 2030. In the last two decades UNICEF has registered 34% decline in global MMR. Between 2000 and 2020, it decreased from 339 deaths per 10,000 live births to 223 deaths per 10,000 live births. The data in Fig. 2.2 shows the MMR trends in India, China and Bangladesh. MMR in far below the SDG target in 2020, while India and Bangladesh reduced the count 384 to 103 and 441 to 123. However, national target of India was to reduce the MMR to 100/100000 live births, country has achieved the target in 2020, by reducing the number to 97/ 100,000 live births (PIB, 2023a, 2023b).

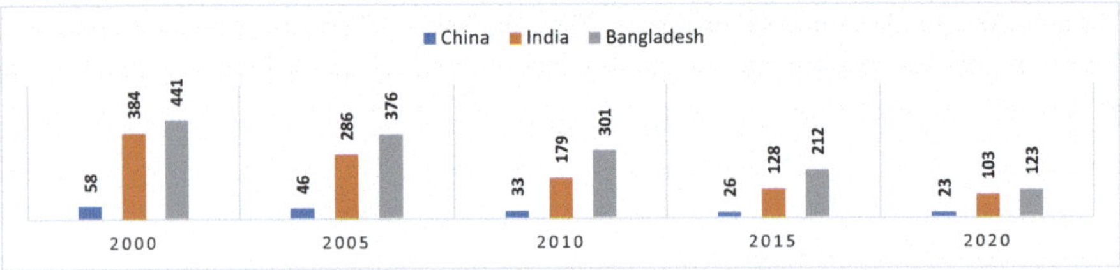

Fig. 2.2 Trend of MMR in China, India and Bangladesh. *Source* UNICEF (2023)

Preventable Deaths Among the Children (Up to 5 Years of Age)

Children are key drivers to the country's present and future. World Health Organization observed that protecting and improving child health is core humanity (WHO, 2023). Nehru recognized this importance when he said *"Children are like buds in a garden and should be carefully and lovingly nurtured, as they are future of the nation and the citizen of tomorrow"* (Mint, 2023a, 2023b). The care of child is an important human duty because children are inherently prone to disease which have an adverse effect on the overall development and wellbeing throughout their life span. An unhealthy child faces many challenges in pursing lifetime opportunities (Chapman et al., 2023).

The leading cause of mortality among children under 5 years of age (U5MR) are infectious disease, pre-term birth complication, birth asphyxia, and abnormality. These diseases can be prevented by lifesaving interventions such as skilled delivery, access to healthcare, healthy diet, and timely treatment. SDG 3.2, aims to bring down the number of preventable deaths of newborn children as well as children under five years of age below 12/1000 births and 25/1000 births respectively. However, the success is not at par with the progress of global MMR.

Figure 2.3 shows the considerable decline in Neo-Natal Mortality Rate (NMR) and U5MR in India. The number declined from 57/1000 births to 19/1000 births and 127/1000 births to 31/1000 births respectively during 1990 and 2021. China has achieved the target of SDG 3.2 while Bangladesh performed better than India. NMR declined from 66//1000 births to 16/1000 births and U5MR 16/1000 births during same period.

SDG4: Quality Education to All

Education is key support to a thriving democracy. It enables citizens in realizing their full potential, makes them full participant in public affairs and societal activities, consequently accelerating nation prosperity. Ensuring equitable quality education to all is fundamental to SDGs. SDG4 aims to promote lifelong learning opportunities to all by 2030 (Saini et al.,2023). Education is vital for development. Studies strongly

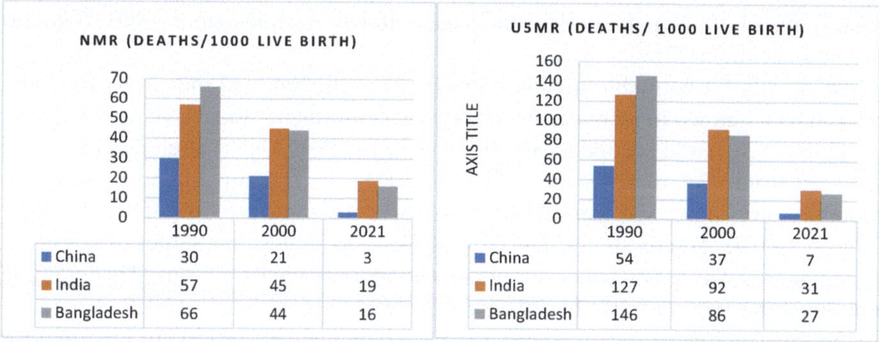

Fig. 2.3 Trend of NMR and U5MR in India. *Source* United Nations International Children's Emergency Fund (2022)

suggest that it is an important tool to uplift people from deprivation (Hannum & Buchmann, 2005; Fomba et al., 2023). Education enhances the capacity to make better decision and help in shaping a better citizen (Sen, 1999). However, developing countries are struggling to improve the educational indicators. India is one among them.

The youth have potential to advance the social and economic prosperity of India, if the country is able to equip them with high quality education. The National Education Policy (NEP) 2020 recognized the causality between universal quality education and human capital. Subsequently, India has restructured the education system in coherence with universal education system to promote and encourage learning, a perquisite to achieve all SDGs. The objective of NEP 2020 is to enhance the unique potential of each individual.

India has prioritized education as a fundamental right and enacted comprehensive policies to ensure quality education for all in consequential legislation envisaged in Article 21 A of Indian Constitution, free and compulsory education as a fundamental right of every child in age group 6–14 years. The Government of India enacted Right to Education Act 2009 to provide free and compulsory education to all children in age of 8–14 years and also reduced the distance of school from house. Mid-day meal encouraged students to stay in schools and helped in improving health of students. Sarva Siksha Abhiyan was dedicated to improving school infrastructure, providing training to teachers, and enrolling out of school students in formal education.

India's growth story includes the stories of educational advancement at all level. The benefit of economic growth is reflected in the delivery of education service and the improvement in the student–teacher ratio at primary level. UNESCO (2020) study noted significant contribution of India in realizing global target of equal enrolment of boys and girls in primary and secondary education in 2018, in comparison to 90 girls per 100 boys in 1995. In India, the number of girls enrolment is higher than boys at both levels. In the last few decades, countries across South Asia registered improvement in their education level. IMF literature shows that the literacy rate in

India, Bangladesh, Nepal and Pakistan have doubled or tripled since 1980s (Garcia-Escribano et al., 2022).

SDG4 calls for equitable quality education for all. National indicators for India under SDG4.1 aimed to increase the adjusted net enrolment ratio (ANER) at elementary (1–8 classes) and secondary (9–10 classes) to 100). ANER at elementary level increased from 92.8 in 2018–19 to 96.5 in 2021–22. The corresponding figure at secondary level increased from 57.6 to 64.7 over the same period (UDISE+, 2022; UDISE+, 2019). Trends shows that India is likely to achieve the target of ANER at elementary level by 2030. However, the secondary level calls for focused programme to attained the target of 100 by 2030. At the sub national level, ANER data show mix figure. The lowest for secondary (41.6) and elementary (80.3) are recorded for Nagaland while Tamil Nadu with 90.4 at secondary and 99.5 at elementary level stood on top.

Every child is unique, so their ability to understand and interest vary. This difference begins in early stages, before the formal schooling. Difference in solving math, understanding language and environmental issue play important role in children education level upgradation. SDG4.1.2 and SDG4.1.3 focus on improving the possibility of correct response of grade 5 students and Grade 8 student based on their learning outcomes in language, Mathematics and EVS for grade 5 and Mathematics, Science and Social science for grade 8. The national target is 67.89 for Grade 5 and 57.17 for grade 8 by 2030. The National Achievement Survey 2021 reveals that performance of pupils at V grade and VIII grade is 58.46% and 53.1% (NAS, 2021). Fifth grade students in urban areas are better in language proficiency in comparison to rural students. More than 60% students performed better at eighth grade, performance in Math science and social science stood at 51%, 50% and 51 respectively. Trend indicates attainment of targets.

Children out of formal education and dropout are a major challenge for India in the realization of SDGs. SDG4.1.4 is committed to reduce the proportion of children out of school in age group 6–13 years to 0.28 and SDG 4.1.5 is dedicated to bring down average annual dropout at secondary level to 10. Data shows that drop rate at secondary level has been and continue to decline from 17.9 in 2018–19 to 12.6 in 2021–22. However, subnational figures are uneven, with 27.3 Odisha at the bottom while Chandigarh and Lakshadweep are at the top with zero cases in 2021–22.

Quality of teachers play a huge role in retaining students in school. A qualified teacher is a source of motivation and inspiration for the students. SDG4.C ensures the supply of good teachers. India targets to equip all the schools with 100% qualified teachers and maintain the pupil-teacher at less than/equal to 30 by 2030. Recent data reveals that at pre-primary, Primary and upper-primary, the percentage of trained teachers are 67.6, 88.7 and 88.6 respectively. The pupil-teacher ratio at national level has improved in primary level from 28 in 2018–19 to 26 in 2021–22 and from 30 to 27 at secondary level during same time.

SDG5: Gender Equality in India

Gender equality is indispensable for realizing a peaceful and sustainable existence. It is a precondition for sustainable development. Throughout history, biases and prejudice against women cut through geographies, culture, and societies that resulted in deep-rooted structures snatching away women's rights and dignity as equal human beings. Gender equality is conceived in Mahatma Gandhi's ideal society, a society that thrives only when women who are characterize as *abla* (helpless) become *sabla* (empowered). Ambedkar also measured the progress of society by the degree of progress achieved by women. Many economists recognized the inextricable link between gender equality and human development. Sen's idea of development goes through "expansion of freedom for all for men and women equally"; Martha identified the need of social and political equality for fostering equal capabilities; Esther Dufflo recognized causality between economic progress and women empowerment. Post- independence, the constitution of India has recognized the principle of gender equality. Article 14 grants equality to women and Article 15 (1) prohibit any biasness based on sex. Further article 15 (3) empowers states to ensure positive discrimination in favor of women. Article 16 ensures unbiased opportunities to all Indians (The Constitution of India, 1950).

SDG 5 aims to promote gender equality in all forms. India has prioritized six indicators out of the nine identified global goals at international level: SDG5.1 (child sex ratio), SDG5.2 (gender pay gap), SDG5.2 (removal of violence against married women by their spouse), SDG5.5.1 (gender equality in decision making at all levels of Governance), SDG5.5.5.2 (equal participation of women in Labor force), SDG5.6.1 (ensures use of modern method of family planning to all women in age group 15–49 years of age).

SEX Ratio at Birth in India

SDG5.1.1 ensures balanced child sex ratio at birth. The natural normal sex ratio at birth ranges from 102–106 girls per 100 boys. Nevertheless, on many occasions, the number of girls is observed higher than 130 per 100 boys (WHO, 2011). However, available data indicates that sex ratio at birth has been increasing since the 1980s. Particularly, the situation in Asian countries like China, India, South Korea, Vietnam are more severe. The endurance of unfavorable sex ratio has been starkly brought in to literature by Amartya Sen when he coined the term 'missing women' to refer to the survival of female child in extreme gender biased in son preference countries (Sen, 1990; Pennington et al., 2023). Increasing sex ratio is credited to human intervention like prenatal sex detection, desire for male child, and so on. The United Nations Populations Funds (UNFPA) reports missing of 142 million around the world (UNFPA, 2023). In 2020, almost seventy-five percent of global missing women were

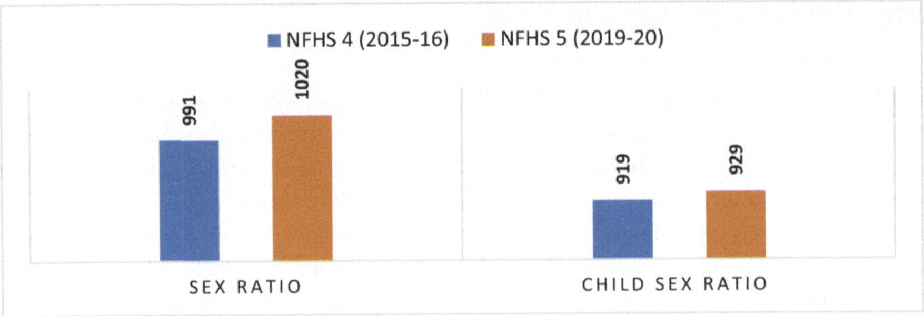

Fig. 2.4 Sex ratio trends in India, NFHS 4 and NFHS 5. *Source* International Institute of Population Studies (2020)

reported from China and India, nearly 72 0.3 million in China and 45.8 million in India (UNFPA, 2020).

Sex ratio in India has been continuously declining in India till 2011 when for the first time in between NFHS 4 (2015–16) and NFHS 5 (2019–21) the country achieved significant rise in sex ratio from 991 to 1020. For the first time girls overnumbered the boys. Over the same period, child sex ratio at birth also registered considerable growth from 919 to 929 (see Fig. 2.4).

However, the progress in child sex ratio is uneven across the states and Union Territories (see Table 2.2). With 817 girls per 1000 boys, Daman & Diu and Dadar & Nagar Haveli emerged at the bottom and Ladak stood at top with 1125 girls per 1000 boys. Further, NFHS 5 reveals that the regions like Chandigarh (838), Goa (838), Kerala (951) Himachal Pradesh (875), Bihar (908) registered a steep decline and are below the national target while Ladak (1125), Sikkim (969), Uttarakhand (984), and Uttar Pradesh (941) shows the considerable growth above the nation average of 929. Nevertheless, Haryana with positive growth of 893 lag behind national and global average (see Fig. 2.3). Uttarakhand, Assam, Lakshadweep, Bihar, Jammu & Kashmir, Kerala Tripura, Andaman & Nicobar and Daman & Diu, Arunachal Pradesh and Ladak have achieved the targets while slow pace of other states are major challenge in realization of target by 2030. Statistics of child sex ratio of all states are provided in Appendix 2.1.

The positive growth in sex ratio and sex ratio at birth is a consequence of strict law against female feticide, awareness and specific programme for lactating mother and new born babies. In 1994, the government of India enacted the Pre-Conception and Pre-Natal Diagnostic Techniques Act (PCPNDT) 1994 to end female feticide through pre-natal sex determination techniques such as ultrasound. Kumar et al. (2014) study contends that in Southern India, parents still prefer sex detection, despite knowing about PCPNDT Act. Ahankari et al. (2015) noted that the awareness and sex determination laws might be significant in improving sex ratio at birth. In January 2015, Government of India launched *Beti Bachao, Beti Padhao* (save daughter, educate daughter) scheme, dedicated to improve sex ratio at birth and ensure social protection to girl child. The key element of the scheme includes promotion and awareness,

Table 2.2 Child Sex Ratio at birth between NFHS 4 and NFHS 5

States/UTs	NFHS 4	NFHS 5	(+) Difference
India	919	929	20
Ladak	823	1125	302
Sikkim	809	969	160
Lakshadweep	905	1051	146
Uttarakhand	888	984	96
Haryana	836	893	57
Uttar Pradesh	903	941	40
States/UTs with maximum decline in Sex ratio at birth			
States/UTs	NFHS 4	NFHS 5	(−) Difference
D&D and DNH	983	817	166
Chandigarh	981	838	143
Goa	966	838	128
Kerala	1047	951	· 96
Himachal Pradesh	923	875	48
Bihar	934	908	26

Source International Institute of Population Studies (2020)

gender sensitization and enforcement of PCPNDT Act. Gupta and et al. (2021), observed positive impact of *Beti Bachao, Beti Pdhao* particularly in Haryana and argued that the effectiveness of scheme depends on implementation and monitoring. Dedicated gender inclusion fund in National Education Policy (NEP) 2020 is a progressive step in advancing the capacity building programme for women in India.

Violence Against Women: Spousal Violence

Violence against women manifest in physical, sexual, emotional and psychological. The World Bank reveals that the most prominent violence ever experienced by women are intimate partner violence and non-partner sexual violence. SDG 5.2 calls to end all forms of violence against women. India has adopted SDG5.2.3 to reduce the proportion of ever married women (15–49 years) experiencing spousal violence to zero by 2030. The World Bank report placed India in the second highest range (35–39%) of spousal violence against women (WHO, 2018). However, NFHS 5 data shows considerable decline in proportion of ever married women in age group of 18–49 years who are subject to spousal violence, from 31.2 in 2015–16 to 29.3 in 2019–21 (IIPS, 2020a, 2020b). In India nearly 1 in 3 women have experienced spousal sexual violence. Section 498 A under Indian Penal Code (IPC) is a legal arrangement to protect married women from their spouse and his relatives. Despite the extensive prevalence of crime against women, it is an underreported crime based on many reasons such as embarrassment, long legal/bureaucratic procedure, financial condition, and many more.

Proportion of Women in National Parliament, State Legislative and Local Bodies

The beauty of democracy is rooted in ensuring equal participation of men and women in the management of public affairs. Full and equal involvement of both men and women in decision making demonstrate the natural composition of society. It also improves efficiency of the political processes by addressing the concern of each and every society in more democratic and effective way. In order to realize democratic values, women should be free from oppression in any forms. For far too long, underrepresentation of women in decision making at all levels of governance pushed them to social, political and economic marginalization. SDG 5.5.1 promotes gender equality in decision making at national parliament and state legislative. It aims to increase the share of women in Parliament and State legislative Assemblies up to fifty percent.

The occupancy of women in local bodies in India is 44% and the corresponding figure for Bangladesh is 23% (UN Women, 2023). The number of elected women in India is around 1.4 million (UNFPA, 2024). The tremendous growth in women representation in local bodies in India and Bangladesh was achieved by ensuring 1/3rd reservation for women in local bodies through Panchayati Raj of Act 1993 in India and the local government (Union Parishad) Second Amendment Act 1997 in Bangladesh. (Khan & Ara, 2006) The Government of India has further amended the Panchayati raj Act 1993 to enhance reservation from 1/3rd to 50%, which takes India on path of equality (PIB, 2011).

Nevertheless, achievement of equality in decision making at third tier has not pushed many women into the second and first tier. Inter-Parliamentary Union data shows the glaring gap in the composition in lower house of Indian parliament (Table 2.3). The share of women is just 15.2%, even below China with 26.5% and Bangladesh (20.9%). In the two most populous country, India and China, women's share in the ministerial position is only 7% and 4% respectively while in Bangladesh it is 10% (IPU, 2023). The hopeless proportion of women in national Parliament and State Legislative could be improved by quota, similar to the implementation at the third tier. Realizing the need of quota, in 2023 the Indian government has taken a progressive step by passing the long pending by 128th amendment which makes arrangement for one-third reservation of seats in National Parliament and State legislatives (PIB, 2023a, 2023b). It gives some hope in the realization of SDG 5.1.1 by 2030.

Table 2.3 Proportion of Women in Parliament and Local Bodies

	Lower House			Ministers	Local bodies
	Total seats	Women	%	%	%
India	538	82	15.2	7	44
Bangladesh	350	73	20.9	10	23
China	2977	790	26.5	4	32

Source International Parliamentary Union (2023), UNWomen (2023)

Gender Pay Gap in India

Over time, women's participation in legislative process has increased in tandem with women's growth in labor force (Iversen & Rosenbluth, 2008). SDG5.2 (gender pay gap) calls for equality in labor force. Equal opportunity in the job market is still a far-fetched dream and remains a major hurdle in achieving SDGs. Globally, women face many challenges in access to job opportunity, limitation in decision making and being characterized in stereotype jobs. Further, this horizonal and vertical division clubbed with unequal distribution of unpaid works in care economy leads to gender pay gap and over representation of female in part-time employments.

ILO (2023a, 2023b, 2023c) estimation finds that 24.9% gap in labor force participation in global economy, less than half of women (47.4%) are employed against 72.3% men. In 2019, women earned only 51 cent in the labor market against each dollar earned by men. The corresponding figure in low and lower-middle income countries is even worse, 33 cents and 29 cents against each dollar (ILO, 2023a, 2023b, 2023c). India has recorded considerable decline in gender pay gap, in 1993–1994, on an average, women in India earned 48% less than men, the gap contracted to 28% in 2018–2019. The achievement in gender pay gap is deaccelerated by COVID-19 which cause a 7% hike in the gender pay gap (ILO, 2022a, 2022b).

For example, in September 2019, average monthly wage of regular employee accounted Rs. 13,912 and Rs. 19,194 for rural and urban male respectively while females earned Rs 12,090 and Rs 15,031 in rural and urban area respectively. Self-employed person earned less than regular employee, Rs. 9661 and Rs. 17,166 for rural and urban males during July- September 2019, females pocketed Rs. 4558 and Rs. 7141; a male casual labor earned Rs 297 and Rs. 365 in rural and urban areas respectively, corresponding figure for female is Rs. 185 and Rs. 190 during July-September 2018 (IFC, 2022).

The low participation of women in labor force finds its roots in the traditional value and level of education among women in Indian workforce (age 15 years and above). In 2015, McKinsey Global Institute projected that India could boost its GDP by $0.7 trillion by adding 68 million more women to the Indian workforce (McKinsey, 2015). According to World Bank statistics gender parity in the Indian workforce could add 1.5% to the existing GDP growth rate. Most studies indicate that if India ensures gender parity in the workforce, GDP growth will increase to 8% by 2030 (Mint, 2023a, 2023b).

Gender gap in job market exist at all levels, the gap increases with decrease in education level. Education plays crucial in wage fixation, renumeration for women with lower level is generally less than highly educated women and men with low education level (OECD, 2017). Annual Periodic Labor Survey (July 2021-June 2022) reveals that 40.4% women are illiterate, 22.5 passed secondary and senior secondar, another 25.7% graduates, and 25.8% post-graduates; considerable growth in comparison to PFLS 2019–20, corresponding data of which are 36.8% not literate, 20.8% above secondary/sr. secondary, 24.1% graduates, and 38.1% post-graduates. Data shows significant increase in share of diploma/certificate holders, from 38.2% to 50.5% during the same period (MoSPI, 2023). Illiteracy enforced women to work in agriculture sector and other low skilled informal job in manufacturing and other non-agriculture sectors. Women's participation in agriculture is the highest with 62.9% and followed by 11.2% in manufacturing sector in 2022.

Women occupied 8.9% of managerial position, 15% pilots in India are female, registering the highest proportion of female to males across the world (Boben & Jain, 2023). Increasing trends of women participation in labor force in developing countries shows causality between employment opportunities and education level (Heath & Jayachandran, 2016).

Illiterate women contribute more in overall women labor force, show downward trends with primary and secondary education and again rise with higher education. Level of education is also an important factor in determining women employability. Deshpande and Kabeer (2024) and Olsen and Mehta (2006) finds U Shaped curve in female educational level and employment opportunity. Downward slope shows the illiterate or women with level of education, women at bottom are engaged in domestic work while the upward slope shows that chances of women employability rise with increase in education level. Other factors include households need, women from poorest cohort have been and continue to be reported more in economic activity (Srivastava & Srivastava, 2010).

The total youth population in India is nearly 370 million, 177 million females over 193 million males (UNFPA India, 2022). Trends based on recent PLFS report suggest that the increase in education level among the women, increase the chances of better job. Nevertheless, Periodic Labor Force Participation (PLFS 2021–22) shows tepid growth in women's labor force participation in India, from 30.3% in 2019–20, 32.5 in 2020–21 and 32.8% in 2021–22, registering a positive trend. However, although the level of education among women in India is increasing, they are still deprived from employment opportunities. This indicates mismatch in demand and supply of skill in job market which needs to be addressed. While Indian women are willing to contribute in their household income and participate in labor force, the lack of skill and social norms siege their mobility. There is a need of extra thrust to ongoing programme under National Skill Development Mission, such as *Pradhan Mantri Kaushal Vikash Yojana* (PMKVY), Deen Dayal Upadhyay Grameen Kaushalya Yojana (DDU-GKY), craftsmanship training scheme and craft instructor training programme.

Modern Method of Family Planning in India

India is the first country in the world to implement family programme in 1952. Since implementation the policy has been reoriented to provide umbrella cover to health aspects such as reduction of infant & maternal mortality and morbidity promotion of reproductive health along with the target of population stabilization. Commitment to 100% use of modern family planning technique by women (15–49 years) is one of the key elements as manifested in objectives of various health documents such as National Population Planning 2000; National Health Policy 2017; MDGs and SDGs. SDG5.6.1, ensures use of modern method of family planning to all women in age group 15–49 years of age. India has committed to achieve the target of 100% by 2030. NFHS 5 shows the significant progress in proportion of currently women in age group 15–49 which has increased from 47.8% in 2015–16 to 56.6% in 2019–21 (IIPS, 2020a, 2020b).

SDG6: Clean Water and Sanitation

Water, sanitation, and hygiene (WASH) are key elements of public health. Lack of access to clean water, unvigilant sanitation and poor hygiene leads to risk of communicable water borne disease like cholera, typhoid, hepatitis A and may more (Kanungo et al., 2021). On many occasions, issue of clean water and hygiene is related to poverty and dense population (Nath, 2003). Problem of clean water, sanitation and hygiene is further linked to human loss. World Bank literature finds that one in every ten deaths in India is linked to poor sanitation (World Bank, 2016). SDG 6 recognizes the importance of clean water and Sanitation. The national targets of India (SDG6.1) is to ensure clean drinking water to all Indians. SDG6.2 aims to provide toilets to each rural household and SDG6.3 to make India open defecation free country. SDG6.3.1.4 provides for the installation of sewage treatment in coherence with sewage generate in urban area and SDG4.2. targets fresh water withdrawal against available fresh water by 70%.

As per Multiple Indicator Survey (2023), only 35.5 dwellings in India have facility of piped water, 61.5% in urban area over 24.8 in rural area during 2020–21. However, the proportion of people using improved source of water is 95% at national level, 97.5% in urban area in comparison to 95% in rural area during the same time. Nevertheless, water availability throughout the same source in stark, 32.9% availability of piped water and 60.1% for improved source of drinking water. No significant improvement was recorded in between NSS 76th and 78th round, from 95.4% in 2018 to 95.7% in 2022. Although, access to improve source water do not necessarily refer to clean or sustainable drinking water source. Improved source of drinking water includes handpumps, boreholes, protected dugs, and springs, they are likely to be immune from outside pollution and from faecal matter in particular (WHO, 2022), but prone to chemical contamination. Throughout India, the presence of arsenic, fluorides, iron, nitrate and high level of salinity is reported in ground water (Dhamija, 2023). The chemically contaminate water is not sustainable for drinking purpose, it is serious threat to human health.

In 2016, World Bank joined India's ambitious movement to provide toilets to each household in rural India. Swachh Bharat Mission Gramin is a dedicated programme to promote cleanliness in rural India. The motive is to ensure access of toilets to each household in rural area. The National Annual Rural Sanitation Survey revealed that 90.1% people are living in open defecation free (ODF) verified villages. Sikkim and Tripura had reached the target of 100%, while and Puducherry 63.6%, Dadar and Nagar Haveli 67.2%, Jammu and Kashmir 67.7 and Bihar 71.1 lags behind the national average (NARSS, 2022).

SDG7: Clean and Affordable Energy

World bank literature linked poverty with energy consumption and called it 'energy poverty' (Indrawati, 2015). Poor people are deprived of access to power, and they remain poor if they stay disconnected. For the poor, night starts with sunset because they are impoverished of artificial light. Addressing energy is key to uplifting the poor form poverty, enhancing air quality, reducing health burden, improving household income, and increasing public participation. SDG 7 identified the role of energy and ensure clean affordable energy to all by 2030. India have adopted 3 indicators: SDG7.1.1 to achieve universal electrification of households, SDG 7.1.2 to ensure clean cooking fuel to each household, and SDG 7.2.1 to expand the share of renewable energy to 40%.

In 2005, the government of India launched the *Rajiv Gandhi Grameen Vidyutikaran Yojana* (RGGVY) the most ambitious programme dedicated to financially supporting states to meet household electrification gap. Ray and Pullabhotla (2023) includes RGGVY as one of the most extensive and successful programs committed to rural electrification in India. In 2018, the government of India announced the historic achievement of universal electrification of villages across India. India has marked significant progress in village electrification. 18,337 villages were electrified between 2015 and 2018. However, the success is not reflected at household level as only 8% (1429) villages have universal access to electricity (REC, 2023). The least number of villages with universal household access to electricity in Tripura (2 villages) followed by Himachal (3 Villages), Jammu & Kashmir (7 villages), and Karnataka (12 villages).

One in every 3 household in India rely on primary sources of energy (firewood, chips and crops residue) for cooking fuel in 2020–21. Nevertheless, rural to urban area divide is stark, 46.5% in rural area over 6.5 in urban cohort, 33.8% at the national level. However, 89% in urban area and 49.4% in rural area used LPG for cooking purpose.

According to data collected by WHO, nearly 4 million human lives were claimed by air pollution with unsafe level of fine particulate matter ($PM_{2.5}$) (WHO, 2021a, 2021b). Children are more prone to air pollution (Chafe & Chowdhury, 2021). India shares quarter of death due to air pollution. Nature sustainability literature has noted the double burden of air pollution among the poorest Indian households. Smoke from the burning of fossil fuel for domestic consumption and also excessive exposure to ambient air pollution contributed by general households' consumption, mainly generated by well-off urban households that have higher per capita consumption emission (Rao et al., 2021). New Delhi, capital of India has consistently ranked among the most polluted cities in the world.

The Government of India recognized the financial constrain of people living below poverty line (BPL) in access to clean cooking fuel for domestic consumption. In 2016, the government launched *Pradhan Mantri Ujwala Yojana* (PMUY), a flagship programme to promote clean cooking fuel to rural household. One time cash assistance of Rs 1600 were provided to beneficiaries under PMUY. The target was to reach 5 crore women member of BPL households. Further, in 2018 7 more categories were added such as SC/ST, PMAY, AAY, tea garden and others. In 2021, Ujjwala 2.0 was launched to add remaining poor household.

India's Economic Survey 2021–22 illuminate the country's significant progress in LPG coverage, from 62% to 99.8 between 01/05/2016 and 01/04/2021 (Ministry of Finance, 2022). The data for LPG consumption as cooking fuel provided by Ministry of Finance and Ministry of Statistics and Implementation (NSSO, 2023) are contradicting each other. It implies that household may have LPG connection due to PMUY but they are still preferring primary source of cooking fuel for domestic consumption.

India is committed to achieve the target of installing 500 GW of renewable energy by 2030, in order to raise the share of renewable energy up to 50% of total energy demand. Currently, India has a total of 168.96 GW including 64.38 GW of Solar Power, 51.79 GW Hydro Power, 42.02 Wind Power and 10.77 GW of Bio power. The government has announced an ambitious plan to add another 250 GW renewable power and has invited bids for installing 50 GW renewable capacity annually, starting from 2023–24 till 2027–28 (Anand, 2023).

SDG8: Decent Jobs

The key driver of poverty is unemployment or under employment. For the poor, the way of improving their standard of living goes through formal employment. The importance of formal employment is the mandate of SDG8 which is dedicated to promote decent work for all, a necessary ingredient to achieve sustained, inclusive and sustainable economic growth.

Decent work is a broad term that cuts through productive and freely chosen employment, guaranteeing rights and social security, freedom to organize and participate in social dialogue, ensure basic condition for health and wellbeing of workers and their families (ILO, Acessed on 12 December 2023). However, ILO statistics revealed that the share of informal employment across the world is more than 90%, nearly 2 billion workers, representing 6 in 10 workers in overall employment, and 5 in 10 workers in non-agriculture sector (ILO, 2023a, 2023b, 2023c).

Rapid economic growth has potential to reduce the rate of unemployment, however, it is not the only factor, it also depends on ability of poor response towards surge in labor in more productive categories of employment. Classical literature on economic growth and development observed that, conventionally, low-income level in an economy reflects the larger share of agriculture sector in output and employment. Further, industrialization ushered expansion of jobs and output in manufacturing sector, consequently the share of agriculture sector falls. Subsequently, at later stage of development, service sector replaces the manufacturing sector, there would be surge in opportunities and output in service sector (Fisher, 1935; Clark, 1940; Lewis, 1954; Kuznets, 1966). Kaldor finds the manufacturing sector as "the engine of economic growth" (Kaldor, 1967). Manufacturing sector successfully led economic development over the past half century (Stiglitz, 2021).

Syrquin (1988) highlights the association of economic growth and structural change; economic growth facilitates economic transformation (Nayyar, 2019). The process follows the sequence of shift in employment and output along with other economic activities, from primary sector to secondary sector and finally to the tertiary sector. Consequently, at the micro level the transition process reflects in individual and households, at some extent shift towards modern and formal economy with improved standard of living.

However, the story of India's economic transformation defies the conventional model of causality between economic growth and structural change. Over the period of three-decade, Indian economy realized a sustained economic growth rate when the country successfully registered an annual growth rate of 8.4% (approx.) between 2003–2004 and 2011–2012. Foregoing high growth rate of Indian economy decoupled from rise in formal employment (Chattaraj, 2016). Like most of the developing countries, Indian economy is characterized by economic dualism, wedged between overwhelmed traditional sector and modern sector (Kesar, 2023). Dualistic economy represented by vast "subsistence" sector (labor intensive), and marginal "capitalist" sector (capital intensive) (Chacaltana et al., 2022). The former marked by characteristic such as low wage, relatively less mechanize, low productive, and so forth, while latter identified by high wage, relatively mechanized, highly productive and so forth.

Economic dualism is one of the biggest challenge in developing countries where maximum workers are employed in informal economy (De Vries et al., 2012; La Porta & Shleifer, 2014a, 2014b; McMillan et al., 2017). The dominant characteristic of informal economy is marked by low wage, low productivity and absence of social safety net (Hammer et al., 2022). The explanation is not complete without the ugly feature of vulnerability (ILO, 2002). ILO data (Fig. 2.5) shows the persistence of informal employment in different economic regions. In 2022, informality in Sub-Saharan Africa and Southern Asia reached to 87.26 and 86.75 respectively while the Esat Asian region shows downward trends.

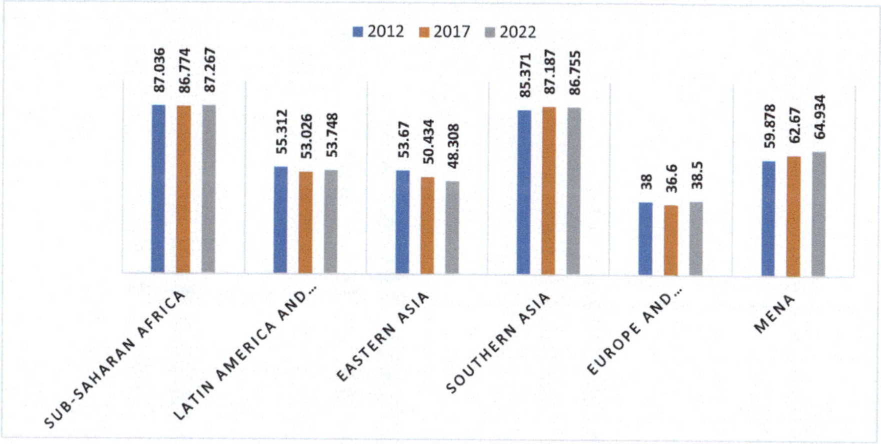

Fig. 2.5 Informal employment Rate (%) in different worlds regions. *Source* International Labor Organization (2022)

Data demonstrate the glaring picture of informal employment (% of total employment) in India, China and Bangladesh within Southern Asia (Fig. 2.6). Despite the sustained high growth rate over the period of three decades, India, and Bangladesh were disconnected with formalization of economy. However, China shows continuous downward trend in share of informal employment, its share reduced to 69.4% in 2018 from 77.5% in 2010, while India and Bangladesh show upward trends.

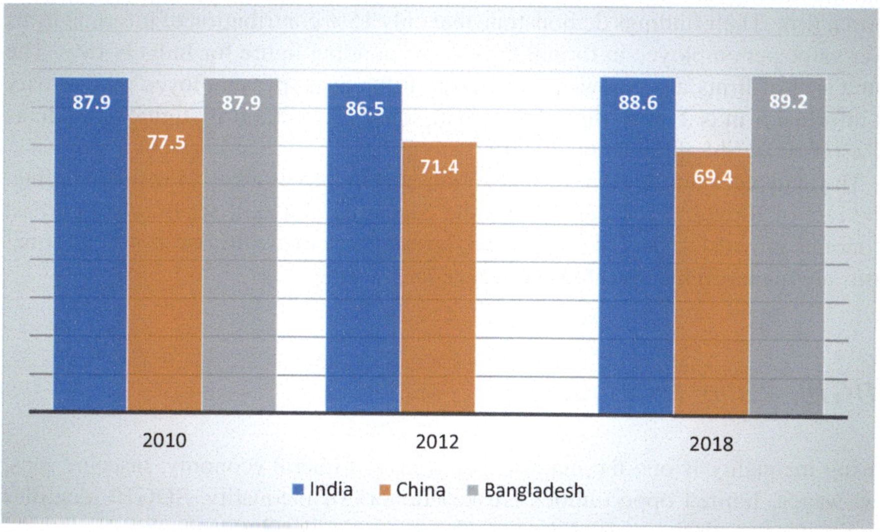

Fig. 2.6 Share of informal employment (% of total employment) in India, China and Bangladesh. *Source* World Bank (2021), Bangladesh Bureau of Statistics (2021)

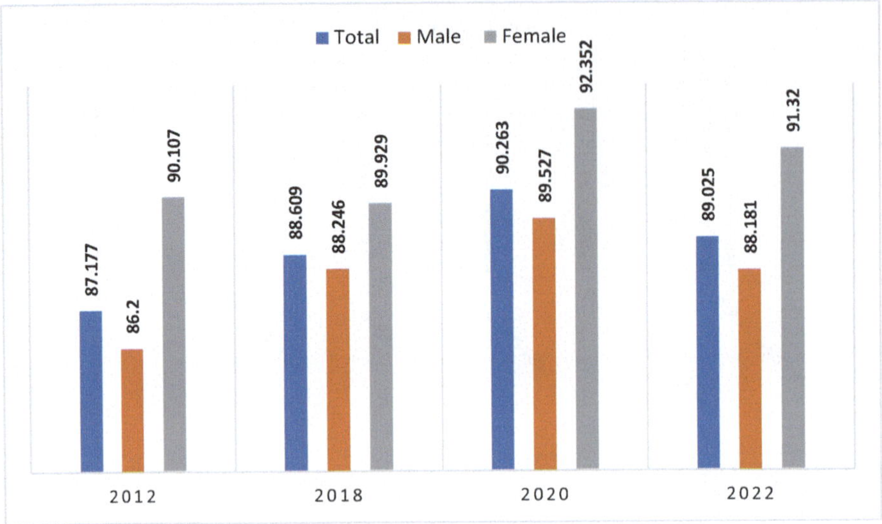

Fig. 2.7 Share of informal in total employment in India. *Source* International Labour Organizations (2022)

The share of informal employment in India is nearly 89%. Data reflects that the proportion of women in informal sector is about 91%, higher than the male share of 88% in 2022 (see Fig. 2.7). During the same period women's share in informality remain almost same or at some extent increasing year by year. Porta & Shleifer attempted to measure the value added by informal firm to the value per employee in formal firm. Their findings demonstrate that only 15% contribution of informal firms over value per employee in formal firm, corresponding figure for India is 18%. The same size of firms also shows productivity differences, per employee productivity of informal firm is 21% of the value per employee of formal firm, figure for India is 21% (Porta & Shleifer, 2014a, 2014b).

The evidences implies that formalization of economy is a key to nation building. The rate of formalization depends on the education and skill set of workers. Fast economic growth has the potential to curb informality only with condition that formal firms are managed by educated entrepreneurs.

SDG10: Rising Inequality

Rising inequality is one the major concern of neo-liberal economy. Insecure jobs, low wages, limited opportunities are key drivers of inequality. SDG10 identifies inequality as a threat to long-term socio-economic development and resolves to reduce inequality. The unprecedented economic growth of India in the twenty-first century had a causal effect on poverty reduction. There is a glaring concern, however, that this growth has been, and continue to be associated to rising inequality (Dang & Lanjouw, 2023). World inequality report 2022 marked India among the most unequal country in the world and soon China may be included in the list (Chancel et al., 2022).

World Inequality Lab observed that the proportion of top 1 person income and wealth reached to 22.6% and 40.1% by 2022–23, highest pick since evolution of Income Tax in British Raj in 1922 (Bharti et al., 2024). The share of top income level during 1950 and 1970, registered sharp decline in share top income level, and faster growth in income level of bottom 50% and 40%F45. Income inequality in India was at par with levels in China in early 1980s, and remained at the same level till the mid-2000s, thereafter China stabilized the pace while it continued in India (Piketty et al., 2019).

The COVID-19 Pandemic and Its Impact in India

The Government of India responded to the pandemic by imposing a national lockdown. On 19th March 2020 India's Prime Minister call for *Janta Curfew (Civil Curfew)* on 22nd of March 2022 from 7 AM to 9 PM. The people who were not involved in essential services were requested to stay at home. We may call it trial before complete nationwide lockdown. Again, on 24th March 2020 Prime Minster of India announced 21 days complete nationwide lock down starting from midnight 24th March till 14th April 2020, which was extended for another 21 days, from 15th April to 3rd May 2020. Further, on 1st May 2020, the Ministry of Home Affairs, Government of India extended lockdown in two phases from 4th May to 17 May 2020 and from 18 May to 31st May (Soni, 2021). In other parts of the world, the United Kingdom announced lockdown "stay at home" on 23rd March 2020 and South Africa entered lockdown on 26th March to till 30 March 2020. The South African approach was different from Indian containment measures which will be discussed in the next chapter.

Vaishnavi Chandrashekhar (2020) reported that 1.3 billion people in India were placed in enforced lockdown due to the pandemic starting from 25th March 2020 onwards. Cross country research of Oxford university found that containment measures in India was the most stringent across the world (Hale et al., 2021). This was the first phase. Swiftly the number of confirmed COVID-19 cases and deaths reached 649,263 and 8166 on 14th September 2020. That was the maximum number since the emergence of pandemic in India (Fig. 2.8). The second wave was much deadlier than the first one. The count of confirmed cases reached to 2,738,957 on 3rd May 2021, subsequently death count reached 28,982 on 31st May 2021.

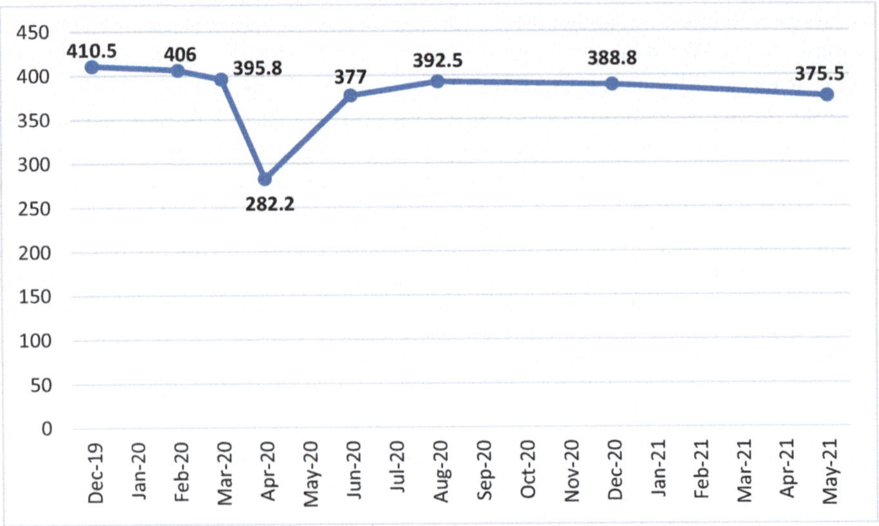

Fig. 2.8 Number of employment (in million) in India. *Source* Centre for Monitoring Indian Economy (2023). *Source* World Health Organizations (2024)

Impact of Pandemic on Indian Economy

India had paid a heavy economic cost for the lockdown (The Economist, 2020). Beyond the health crisis, the lockdown adversely affected the social, economic, health and phycological wellbeing particularly of the marginalized people (Napier-Raman et al., 2021). It was the worst phase in modern history after the great depression (Ram & Yadav, 2021). It disrupted the demand and supply side of the economy as it ceased the firm's production and as well as consumer's consumption. In the early stage of the pandemic, on the supply side, containment measure and individual decision of a worker led to dramatic decrease in production. On demand side, public health precautions and high uncertainty resulted in decrease in total real income.

In India, the lockdown bluntly unmasked the severity of crisis as several industries were shut down, leaving millions of migrant workers stranded. A study conducted by the Azim Premji University (2021) indicated that nearly 80% workers in urban areas reported job loss while 50% informed loss in income or salary delay. The Centre for Monitoring Indian Economy (CMIE) data shows that in December 2019 the number of employed people was 410.5 million (see Fig. 2.8). The number start decreasing in the course of the pandemic. By April 2020, nearly 128 million people had lost their jobs. However, the employment level start increasing with ease in lockdown but sadly it did not reach pre-pandemic level till May 2021 (Vyas, 2023). Lockdown has exposed the vulnerability of informal workers (Dang et al., 2021).

Covid -19 has brought unprecedented challenges to policy makers. Uplifting people from poverty, containing hunger, creating new jobs and bringing back students to their schools is a monumental task and needs immediate government interventions.

To undo the negative setbacks of the covid catastrophes, it is important to empower people on the ground by boosting local capacity. Governments at all levels must collaborate to enhance the efficiency and effectiveness of social infrastructure in general and public health system in particular. The social and economic issues such as poverty, hunger, health, education, gender equality, clean drinking water and sanitation, clean fuel and employment are intricately linked, so any positive change in one is bound to rub-off effects in other domains and subsequently improve the quality of life. The collective approach from supply as well as demand side is essential to ensure overall well-being and to mitigate the impact of devastation caused by the Covid-19 pandemic.

Government Intervention: Stimulus Package to Reignite the Indian Economy

Amid the Pandemic the IMF (2020) projected −3% of global growth rate based on containment measures and multiple crisis (Health Crisis, a financial crisis, contraction of demand and others) faced by the many countries in the first three months of the Pandemic. This translated the containment measure "Lockdown" into "Great Lock Down" as the world over witnessed great recession not seen since the Great Depression. Many policy makers turned to the remedy taken during "Great Depression" i.e. the Keynesian theory. Keynes claimed that Government intervention through policy framework is important to run the market. In the time of crisis Government must solve the problem by immediate intervention rather than wait for an invisible hand to fix the market, as he said "In long run we all are dead". IMF literature noted the Keynes argument that the GDP of an economy is the accumulation of four components: consumption, investment, Government purchase, and net export. Demand of any economy fall if any one of the components fails (Jahan & Mahmud, 2015). Here the role of state is to boost the market by increase spending. In between recession of 2008, Mankiw (2008) suggested that we can stabilize our economies by revisiting Keynes ideas.

Stimulus Package

To meet the immediate challenges of COVID 19, Government of India has given a call for *Atma Nirbhar Bharat* or Self-Reliant India Movement. The foundation of this movement is based on the five pillars; Economy, Infrastructure, System, Vibrant Democracy and Demand. To realize the dream of self-reliant, the Indian Government announced 20 lakh crore packages to stimulate the Indian economy. The financial stimulus package amount was equal to 10% of India's Gross Domestic Product (GDP) to accelerate the revival of various sector such as cottage industry, Micro Small and

Table 2.4 Breakup of 20-lakh crore stimulus package

Items		Amount in crore Rupee
Tranche 1	*Business Including MSME	5, 94,550
Tranche 2	Poor Including Migrants and Farmers	3,10,000
Tranche 3	Agriculture	1,50,000
Tranche 4	New Horizon of Growth	8,100
Tranche 5	Government reform and enablers	40,000
Sub Total		1,10,2650
RBI measure		8, 01,603
Pre PM-Announcement		1,92,800
Total		20,97,053

Source Atma Nirbhar Bharat

Medium Enterprises (MSME), middle class and industries of another sector. This research analyses the various sector of stimulus package, Government policies and the challenges in attainment of SDGs. Breakdown of 20 lakh stimulus package is given in Table 2.4.

India imposed lockdown with effect from 25th March 2020 as a containment measure. The *Pradhan Mantri Garib Kalyan Yojana* (Prime Minister Poor Relief Scheme) was a measure taken by the Government to reduce the distress caused by pandemic and make them food secure. PMGKY is a step toward Zero Hunger (SDG1). The Government sanctioned Rs. 1.70 lakh crore under the said scheme to enable the poor to survive during lockdown. This scheme provides Rs. 50 Lakh insurance coverage to heath workers, 5 kg wheat/rice/person and 1 kg pulses/household through PDS. Apart from food grain supply, Rs.500/month was transferred into 20 crore women *Jandhan* account holder in for the first three months of lockdown.

First Tranche

The first trench of economic relief plan was dedicated to Micro Small and Medium Enterprises (MSMEs). On can understand the importance of MSME when the General Assembly of the United Nations dedicated 27th June as MSME Day. The motive is the promotion of MSMEs contribution in sustainable, inclusive, and resilient economic growth and shared prosperity and decent work for all. Worldwide formal and informal MSMEs hold 90% of world's business and provide 60–70% of employment globally and constitute 50% global economy (World Bank, 2019a, 2019b). Out of the sixteen provisions, six were dedicated to MSMEs. Government injects 3 lakh crores collateral free loans to promote the stressed MSMEs and Rs. 20,000 crores as subordinate debt to MSMEs. Through MSME funds, equity worth

Rs. 50,000 would be infused. Table 2.5 shows the economic relief to the MSME sector with other business.

MSME is key to achieve Sustainable Development Goals. MSMEs are the largest job provider in non-farm sector, second only to agriculture. This sector creates job opportunities that take people out of poverty. It promotes inclusive development by providing job to marginalized section of the society. According to World Bank statistics, most formal employment in emerging economies are created by MSME's, which generate every 7 out of 10 employment (World Bank, 2019a, 2019b).

United Nations agency Department of Economic and Social Affairs report 2020–21 examined the role of MSMEs to each of the SDGs. The report established that the growth in MSMEs sector directly affect the indicators of SDGs such as generating jobs, raising incomes especially for poor and marginalized section of society and providing basic necessities like education, health, water and sanitation (United Nations, 2021). Research indicates that MSMEs directly or indirectly help in realizing the SDGs (Pangarso et al., 2022; Dasaraju & Tambunan, 2023). MSMEs contract poverty (SDG1, No Poverty) by creating job opportunities for the working population. Majority of food producer are engaged in MSMEs which can assure food security leading to SDG2 (Zero Hunger). MSMEs absorb the working age population equipped with SDG4 (Quality Education). It promotes SDG5 (Gender Equality) by providing job to women and girls in large numbers and play significant role in increasing women's effective participation in trade and business. Evidence shows

Table 2.5 Provision Under Business Including MSME

Sr. No	
1	Rs 3 lakh crores Collateral Loans for Business free Automatic Loans, Including MSME
2	Rs 20,000 crore Subordinate Debt for MSMEs
3	Rs 50,000 crore Equity infusion Through MSME Fund of Funds
4	New Definition of MSME
5	Global tender (up to 200 crore) to be disallowed
6	Other intervention for MSME
7	Rs 2500 Crores EPF support for Business & Workers for 3 more months
8	EPF contribution reduced for Business & Workers by Rs 6750 Crore
9	Rs 30,000 crores Liquidity Facility for NBFC/HCs/MFIs
10	Rs 45,000 crore Partial Credit Guarantee Scheme 2.0 for NBFC
11	Rs 90,000 crore Liquidity injection for DISCOMs
12	Relief to Contractors
13	Extension of registration and completion date of Real Estate Projects under RERA
14	Rs 50,000 crore Liquidity through TDS/TCS reduction
15	Other Direct Measures
16	Other Direct Measures

Source Atmanirbhar Bhara (2020)

that MSMEs are successfully bridging the gender gap and that almost 40% of small businesses are owned by women. MSMEs are very helpful in providing decent work and economic growth since it provides platform for technology upgradation and innovation and thus realizing in SDG8 (Decent Work and Economic Growth). MSMEs propel industrialization by providing industrial infrastructure base to rural areas, it promotes innovation with low-cost investment in order to attain SDG9 (Industry, Innovation and Infrastructure). Further increase in income level with decent job in MSMEs for the down trodden population empower them to walk out of poverty thus achieving SDG10 (Reduced Inequalities). Small units are more resilient than big industries which leads them to SDG12 (Responsible Consumption and Production). Finally, the key in realizing the SDG goals lie in learning from the expertise of each other and the public sector must facilitate the private sector with liberal framework to cementing the partnership for development i.e. SDG17 (Partnership for the Goals).

Unfortunately, COVID 19 has put the MSMEs on standby. The Pandemic created the worst recession since the Great Depression. The degree of economic collapse was unprecedented. The crisis resulted in financial instability, together with informal economy, especially in developing countries of Latin America, Sub-Saharan Africa (IMF, 2021). World Bank (2020) estimates shows that the world's economy shrunk by 5.2% in the first year of the pandemic and its Global Economic Prospect reported that the situation represents the deepest recession since the Second World War and the sharpest decline in per capita output since 1870. India was not untouched. Indian economy experienced a devastating contraction of 23.9% in quarter ending in June 2020. This contraction was the sharpest amongst all the major world economies (Slater, 2020). Its effect resulted in low demand and production; World Bank (2020) stated that this situation translated into exodus of migrants from the urban areas.

Status of Micro, Small and Medium Enterprises (MSME) in Indian Economy

MSME sector was not untouched by the misery of the Pandemic. The sector which employs 1100 lakh workers, MSMEs are the second largest employment provider after Agriculture in India. It helped India in removing poverty by creating employment and entrepreneurship opportunities on a large scale. Studies suggest that the employment growth is directly dependent on the performance of Small-Scale Industries (Ayyagari et al., 2014; Delfmann & Koster, 2016; Khanna et al., 2021). In terms of production capacity, small industries with low investment rate perform better than others (De & Nagaraj, 2014), which was reflected in MSME Annual Report (2022). As per the report the contribution of MSME sector in India's Gross value Addition (GVA) and Gross Domestic Product (GDP) contribution is 12.88% and 30.27% respectively at current price from 2014–15 to 2018–19.

In the manufacturing sector of Indian Economy, total contribution of MSME in total manufacturing GVO (Output) at current price remained constant at 33% (One third) during 2014–2018.

MSME performance in India has been remarkable (Table 2.6). The National Sample Survey (NSS) 73rd round found that there were 633.88 lakh unincorporated MSMEs in non-agriculture sector engaged in various economic activities such as manufacturing (196.65) Non-Captive Electric Generation and Transmission (0.03 Lakh), Trade (230.35 lakh) and in others service (206.85 lakh). Annual Report of Micro, Small, and Medium Enterprises shows the activity wise division of MSMEs in India. As show in Fig. 2.9, 36% of MSME are engaged in trade, followed by 33% in other services, 31% in manufacturing and 0.03% in electricity.

Table 2.6 Trends of MSME contribution in Gross value Addition (GVA) and Gross Domestic Product (GDP)

Figure in rupees crore

Year	Total GVA	Growth	Total GVA	Share of MSME In GVA	India's total GDP	Contribution of MSME in India's GDP
2014–15	3,658,196	–	11,504,279	31.80	12,467,959	29.34
2015–16	4,059,660	10.97	12,574,499	32.28	13,771,874	29.48
2016–17	4,502,129	10.90	13,965,200	32.24	15,391,669	29.25
2017–18	5,086,493	12.98	15,513,122	32.79	17,098,304	29.75
2018–19	5,741,765	12.88	17,139,962	33.50	18,971,237	30.27

Source Central Statistics office (CSO), Ministry of Statistics and Programme Implementation

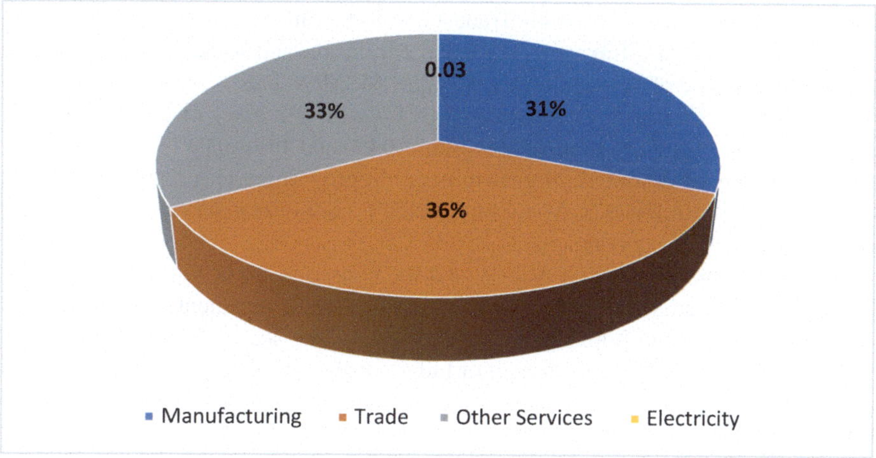

Fig. 2.9 Activity Wise Division of MSMEs in India (%). *Source* Ministry of Micro, Small, and Medium Enterprises (MMSME), GoI, (2022)

Almost 50% (324.88 lakh out of total 633.88 MSMEs) are located in rural India. With a strong pan-India presence MSMEs engaged 111 million workers, 38.718 million in trade, 36.282 million in other services, 36.04 million in manufacturing, and 0.0007 million in electricity. MSMEs contributed around one third of India's GDP, 45% manufacturing output and 48% share in export (MSME, Government of India, 2021). Small enterprises in India were hard hit by Covid-19 (Behera et al., 2020). Their finding suggest that the mean capacity utilization fell flat from 75% to mere 13% and further more alarming that 54% were producing zero unit. The Government of India has taken cognizance of the miserable conditions of a key sector of Indian economy which has potential to reverse the impact of COVID 19 and take the country on path of SDGs.

The Government of India announced several support restructures programs to support MSMEs on a regular interval. In the wake of the Pandemic, government revised the criteria for classification of MSME (Table 2.4). Under the revised criteria, manufacturing and service unit having Rs 1crore of investment with annual turnover of 5 crore is classified as 'micro' while unit venture of Rs 10 crore with annual turnover of Rs. 50 crore is termed as 'small' whereas a unit with Rs 50 crore investment and Rs 250 crore turnover is categorized as 'medium' sector enterprise. Manufacturing and service sector were unified by new composite formula of classification. To promote export, the new provisions excluded export from account of annual turnover. To strengthen the MSMEs, global tender up to Rs. 200 crores were disallowed.

Other Business

Pandemic has changed the world. Trade fairs and exhibition were prohibited and Government provided e-linkages for virtual exhibition and trade fair. Government has a set time frame of 45 days for disbursal of MSMEs dues.

Non-Banking Financial Companies (NBFC) with 9% asset of Indian financial sector positioned third after Schedules Bank (64% and Insurance (14%). In recent years NBFC scored double digit growth in credit providing. This distinctly reflects the borrower faith in NBFCs. Many of the NBFCs are lender serving the borrower from the bottom of the pyramids.

NBFCs were hard hit by COVID19 and the consequence of liquidity constraint of corporate and real sector made the situation worse. Government stepped up with Rs 30,000 crore special liquidity packages to revive the sector. Securities and loan were fully guaranteed by Government of India. Further another Rs 45,000 crore was injected through Partial Credit Guarantee Scheme (PCGS) to do fresh lending to MSMEs with Government guarantee to bear 20% of first loss.

Power distribution Company DISCOM is a loss-making entity of Government sector depending on Government subsidy. The downturn of economy augmented the outstanding dues to Rs 1.39 lakh till October 2020. There was an increment of 34% in generators dues in the same period. To save the DISCOM and relief for immediate debt repayment to generation companies Government of India infused Rs90000 crore.

To strengthen infrastructure, Government of India provided 6-month extension to contractors without any additional cost. The Government released Rs 50,000 crore liquidity by reducing the rate of tax deduction by 25% of existing rate. Government has provided various tax relief measures. Tax return filling and tax audit due date were extended.

Second Tranche

National lockdown has thrown hundreds of thousands of people out of work. The consequences of lockdown were harsh for poor and migrant workers as their source of income dried. The second trench of economic relief package were dedicated especially to migrant workers and the urban poor. The provision under second recommendation is depicted in Table 2.7.

Stringent lockdown measure forced the poor/daily wage workers/migrants' laborer to remain in their house. They needed food assistance as they were not allowed to go outside of their dwelling. The first three provision are meant for migrant labor and for which Government approved Rs 3500 crore to provide them 5 kg of grains and 1 kg of pulses free of cost for two months. 8 crore migrants were estimated to be beneficiaries of this scheme. To resolve the misery of migrant laborers, Government also introduced the technology enabled One Nation one Card to avail the benefit of Public Distribution System (PDS) throughout India. Incentives were provided to manufacturing units, institutions, associations to promote them to build Affordable Rental Housing Complexes on their private land.

Table 2.7 Provision Poor and Migrant workers

Sr no	
1	Free food grain supply to migrants
2	Technology enabled One Nation One Card
3	Affordable rental housing complexes (ARHC) for migrant's worker/Urban poor
4	Interest subvention support of 2% for MUDRA Shishu loan payee
5	Rs 5000crore special credit facility for street vendors
6	Rs.70000 core for extension of CLSS
8	Rs 6000 crore employment push using CAMPA fund
9	Rs 30,000 crore additional emergency working capital fund for farmers through NABARD
10	Rs 2 lakh crore of concessional credit boost to 2.5crore farmers through kisan credit card

Source PIB, 14th May 2020

Like migrants the economy of street vendors collapsed too. Street vendors constitute 2% of the population in cities in India. They constitute an important segment of India's unorganized sector (SEWA, 2014). To reverse the adverse impact of the Pandemic, the Government provided special credit facility to street vendors. In order to help 50 lakh street vendor Government sanctioned Rs 5000 crore. Government also provided interest subvention of 2% for the MUDRA shishu loan (loan amount < Rs 50,000). Government infused Rs 1500 crore to small entrepreneurs.

Another affected group was the middle-income group. COVID 19 pushed 75 million middle income group into poverty and the size of middle class shrunk to 32 million (Kochhar, 2021). This group can be further divided into Government employee, employee in private sector with future security and another category consisting of shopkeepers' sector and those engaged in allied sector like transportation, catering, hospitality or entertainment. The later section was devasted by pandemic as most of them lived in rented houses. To provide them affordable accommodation, Government injected Rs 70,000 crore through Credit Linked Subsidy Scheme for such group.

Centre for Monitoring Indian Economy reported India had lost 7 million jobs due to COVID 19 between February 2020 and February 2021 (Vyas, 2021). Job creation became the prime concern for the Government. Government injected Rs 6000 crore through Compensatory Afforestation Fund Act 2016 to be utilized by state Government to create job opportunities. To facilitate small and marginalized farmers, Government also infused Rs 30,000 crore in addition to Rs 90,000 crore through NABARAD normal refinance route to Reginal Rural Banks & Co-operative Banks. Another sector of farmers involved in fisheries and animal husbandry were provided Rs 2 lakh crore relief package for 2.5 crore beneficiaries.

Third Tranche

India is the leading producer of milk, pulses and jute and the second largest producer of rice wheat, sugarcane, groundnut, and vegetables. Agriculture is the biggest source of livelihood in India. Most of the people involved in agriculture live in rural areas. The pandemic exposed the pathetic condition of rural India. Rural workforce is not salaried as the majority are daily wagers, joint report of Azim Premji University and CII includes that 85% of rural and 53% of urban workforce are engaged in non-salaried jobs (Azim Premji University & CII, 2021). Self-employment and dependency on casual work signaled that almost 90% of India's household witnessed income reduction because of containment measure. Farmers felt acute distress too as the lockdown washed their expectation of good earning from the bumper production of rabi season. Third trench of economic stimulus package deals with growth of agriculture and allied sector, ranging from infrastructure to reform as reflected in Table 2.8.

Notwithstanding India's position as the leading producer of fruits and vegetable, India wasted 40 to 50% of total production of perishable fruits, vegetable and milk

Table 2.8 Provision for Agriculture and Allied Sector

Sr no	
1	Rs 1 Lakh crore for Agriculture Infrastructure
2	Rs 10,000 crore to Formalization Plan of Micro Food Enterprises
3	Rs 20,000 to PMMSY for fisherman
4	Rs 15,000 crore for Infrastructure Development of Animan Husbandry
5	Rs 13,343 crore for National Animal Disease Programme
6	Rs 4000 for Herbal Cultivation Promotion
7	Rs 500 crore for Beekeeping
8	Rs 500 crore for TOP-to-TOTAL

Source AtmanirbharBhara (2020)

valued at USD 440 billion due to the lack of cold storage. The total capacity of Indian cold storage is 30.11 million metric ton in about 6300 cold storage which accounts only for a mere 11% of total production of perishable food (India Today, 2017). The severe situation demanded immediate intervention. Government of India attempted to develop agriculture infrastructure by injecting Rs 1 lakh crore to make agriculture profitable for farmers including availability of cold storage and post-harvest management. For the allied sector, Rs 10,000 crore was sanctioned to formalize Micro food enterprises.

The potential of aquaculture in India is yet to be fully realized. 2.36 million ha of pond and tanks 1.24 million ha of backwater provide a great opportunity of employment in rural areas, particularly in transportation and retailing. To tap the potential of fisheries, the Government approved Rs 20,000 crore through Pradhan Mantri Matasya Sampada Yojana. The fund was distributed in two parts, Rs 11,000 crore for inland aquaculture engagement and Rs 9000 crore for infrastructure development. To protect animals from mouth and foot diseases Government invested Rs 13,343 crores. This scheme benefited 53 crore animals (cattle, buffalo, sheep, goat and pig) with 100% vaccination. Further, Rs 1500 was approved for animal husbandry infrastructure.

India has a large base of herbal industry which represents 8610 herbal units and thousands of unregulated herbal units with millions of traditional naturopaths. Marketing of these medicinal plant are done through mandis. To develop the network of mandis at regional level and bring more land under cultivation of medicinal plant, Rs 4000 crore was sanctioned to promote the sector. Apart from this Government is also set to spend Rs 500 crore on Bee keeping farming.

Government has allocated Rs 500 crore to reduce spoilage and facilitate the delivery of tomatoes, onions and potatoes (TOP) along with other fruits and vegetables (TOTAL). To make agriculture sector competitive and enable farmers to attract investment, the Government also amended the essential commodity Act 1955.

Fourth Tranche

The motive of *Atma Nirbhar Bharat* (Self-reliant India) is to make India self-reliant and prepare the country for global competitiveness. To increase the participation of people in many sectors, Government also simplified the norms, bought transparency and provided necessary boost for growth and employment. The sectors in which these are given in Table 2.9.

The fourth trench was all about structural reform of public sector establishment. Government had liberalized the norm to attract investment. Public sector banks were reformed to promote professionalism and many other steps were taken to create more job opportunities by facilitating "ease of doing business in India".

First and foremost, the coal sector was revitalized. Coal sector had been a much-neglected sector. Even during the colonial period, the British never tried to mechanize the coal sector in India because of availability of cheap labor in India. British India imported coals from Britain to fuel the railways in western India. Although, India is the world's third largest in coal production and fifth largest in term of coal deposit, the country still import coal. Therefore, to tap the available untapped coal mines, the Government decided to commercialize the sector by removing Government monopoly in the coal sector. Commercialization has opened the market for all and any firm can bid for coal and sell in the competitive market.

To enhance the efficiency of Mining sector, the Government has removed the distinction criteria between captive and non-captive mine and provided for the transfer of lease and sale of surplus minerals. Seamless exploration-cum-mining-cum-production regime was introduced to restructure the mining sector targeted to auction 500 mining blocks. It has further been proposed to join auction Coal and bauxite block to improve the completeness of aluminum industry by reducing the electricity cost.

In the defense sector, Government has raised the FDI cap from 49 to 75%. To reduce the dependency on import in defense sector, Government has prepared the list of articles banned for import. The list will expand continuously in future and the banned articles are required to be produced and purchased within the country.

Table 2.9 Various sectors transformed under 4th trench

Sr no	Sectors
1	Coal
2	Minerals
3	Defense Production
4	Civil Aviation
5	Power Distribution
6	Social Infrastructure
7	Space
8	Atomic Energy

Source AtmanirbharBhara (2020)

To reduce the burden of heavy fare in aviation sector, Government has proposed to ease the restriction from 60% unutilized India's airspace. This initiative makes aviation sector richer by Rs 100 crore/year.

The poor state of Power distribution company DISCOM is sought to be improved by privatization. DISCOM of Union Territories should be privatized. Other reforms related to consumer right, industry's promotion and suitability has been done.

To meet the immediate requirement in social infrastructure projects, Government has made provisions of Rs 81,000 crore by the route of Viability Gap Funding (VGF). Centre had increased its share in central sponsored projects from 20 to 30% as VGA.

To expand the space sector Government announced its willingness to invite private companies to create a level playing field in satellite launches and space-based services. Government liberalized the geo-spatial data policy to provide its benefits to tech-entrepreneur.

In atomic sector Government has allowed PPP model in establishment of research reactor to produce medical isotopes use in the treatment of cancer and other dieses. Through PPP mode Government enhanced the irradiation technology for preservation, a step which also assist farmers in preserving their produce.

Fifth Tranche

Fifth trench advocate structural reforms such as excluding debt default due to Pandemic under insolvency and bankruptcy code (IBC) and modifications related to surge in state borrowing limits and additional fund of Rs 40,000 granted to Mahatam Gandhi National Rural Employment Guarantee Act (MGNREGA).

Mahatma Gandhi National Rural Employment Guarantee Act 2005

During the lock down India witnessed severe challenges of unemployment and reverse migration. In the absence of social security migrant workers began to move back to their native places in hope to sustain them better than the host locality. Huge proportions of these laborers are daily wage earners and the absence of work due to the pandemic created panic among them. The unprecedented circumstances made it difficult to afford the expensive life of cities. As per census 2011, there were 456 million internal migrants, out of whom 194 million are workers (Srivastava, 2020). As per data shared by Mininstry of Labour and Employment, Government of India, more than 1.14 crore migrant workers returned to their homes (Nath, 2021). The unemployment rate reached to historical high level at 23.5% in March 2020 (see Fig. 2.10) (Rathore, 2023).

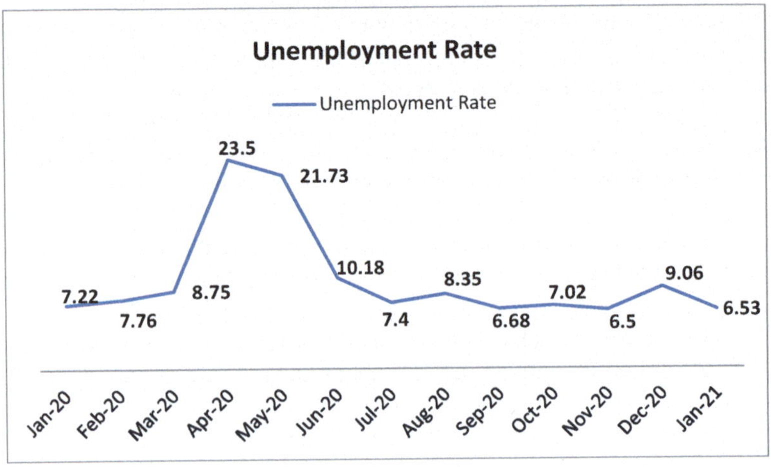

Fig. 2.10 Trends of unemployment During COVID 19 (January 2020-January 2021). *Source* STATISTA (2023)

Rural area was less affected by the COVID 19 as compared to urban areas. The influx of mass population to rural areas rendered the situation miserable. Still there was some hope in rural areas because agriculture and allied activities were exempted from lockdown and a succor scheme Mahatma Gandhi National Rural Guarantee Act (MGNREGA) provided jobs in the tough time of pandemic. Conceived to relive from the burden of unemployment and its byproduct hunger in rural India, the MGNREGA is the world's largest social security and public work programme providing a minimum of 100 days job for the protection of livelihood of the household in rural India. This Scheme was enacted in 2005 for the upliftment of India's rural population by providing 100 days job in every financial year. MGNREGA has played a crucial role in sustaining livelihood of informal workers. MGNREGA is an Act rooted in Indian Constitution providing entitlement of 100 days of job. It goes beyond the notion of social safety net and ensures employment as right (Mitra, 2022; Holmes et al., 2011). It is the only flagship programme run by Government of India for the promotion of wage employment in rural India (Turangi, 2021). Since the MGNREGA is framed by central government, the center bore the larger part of expenditure. The responsibility of implementation is entrusted to the state government.

The existing studies show that MGNREGA has a sound record of inclusive development in rural India. Dreze, Dey, & Khera (Dreze, Dey, & Khera, 2008) in their studies found the performance of MGNREGA to be far better than any employment scheme run by central and state Governments. Jacob (2013) examined the potential of MGNREGA in meeting the rural–urban gap regarding social infrastructure such roads, sanitation and in raising agriculture productivity by strengthening irrigation systems. It creates a stable earning source for the marginalized population resulting in a smoother income graph in between 100 days job and agriculture season. Khera (2008) showed that MGNREGA work performance improves with involvement of farmers' organisation. Khera and Nayak (2009) conducted a study on participation of women workers in respect of MGNREGA. Their study surveyed the kaimur and Araria district of Bihar, Surguja of Chhattisgarh, Palamu and Kodarma of Jharkhand Badwani and sidhi of Madhya Pradesh, Dungarpur and Sirohi of Rajasthan and Sitapur of Uttar Pradesh to find out the realization of MGNREGA's potential. The engagement of women in MGNAREGA varies from state-to-state. The overall women participation rate was estimated at 32% while the participation of women in Rajasthan reached 71%. Adhikari and Bhatia (2010) argued that direct cash transfer into workers bank's account curbed corruption practices. Azam (2012) shows the positive impact of MGNREGA on agriculture wages. Liu and Deininger and Liu (2013a, 2013b); Ravi and Engler (2015) argued that the programme raises the consumption expenditure, energy intake and asset creation. Afridi et. al., (2016) found that women's earning through MGNREGA is reflected in better educational attainment of children. Related studies showed the positive impact of MGNREGA on inclusive growth in rural India by providing social safety net to livelihood security, and democratic governance (Misra, 2019; Patwardhan & Tasciotti, 2022). Sharma, Saluja, and Sarma (2016) argued that increment in purchasing power of MGNREGA beneficiaries has expanded the domestic market.

Taking cognizance of the importance of MGNREGA for India's rural population, Government of India's Finance Minister allocated an additional fund Rs. 40,000 crores to the estimated budget of Rs. 61,000 crores to accelerate the growth in rural area by creating 300 person days job through MGNAREGA. The role of MGNREGA in providing employment to unskilled workforce, including jobless migrant workers who returned home due to uncertainty caused by pandemic (The Telegraph, 2021). Study at Azim Premji University (2022) noted that the work demand under MGNREGA continued to increase during 2021–2022, nevertheless central government bring down the budget allocations to pre pandemic level in following years. Very recent ILO literature includes that MGNREGA act has significantly contributed in rise of wages in rural workers (Ananian & Dellaferrera, 2024).

Challenges

A major challenge in achieving the SDGs by 2030 is the mixed progress across the states. Some states are performing better than the national average and many are lagging behind. Populous states like Bihar (33.76%), Jharkhand (28.81%), Uttar Pradesh (22.9%) and Madhya Pradesh (20.63%) are in the lower rung, on other side Kerala (0.55%), Goa (0.84%) have already achieved the targets of no poverty. The states with highest number of poor also lead in outside migration in search of better opportunities. Most of them get into low skill job.

Employment is at the core of poverty reduction. There are multiple factors that keep people unemployed or out of active labor force. Figure 3.11 shows that the number of employed persons has not returned to the pre pandemic level. However, while majority of men have returned to work, the same recovery is not reflected among the women workers who lost their job due to lockdown (Abraham et al., 2021). Nearly 90% of Indian workforce are employed in under-productive informal sectors, living without any social safety net.

India is a country of youth and the biggest challenge is providing quality education to the world's largest youth population. The Centre for Monitoring Indian Economy data shows that unemployment among youth is at all-time high, 45.4% of India's youth (15–24 years) population are unemployed (Somayya, 2023). As shown in Fig. 2.11, the youth's labor force participation continues to be on the decline. Skill India Report (2023) reports indicates that only half of the young Indians are employable. There is an urgent need to address youth unemployment.

To overcome the shortage of food during COVID- 19 pandemic, the government launched Pradhan Mantri Garib Kalyan Yojana, a dole to make the marginalized food secure and the technology enabled *adhar* card to avail the benefit of scheme. The bigger challenge to realize the stimulus package is in its implementation on the ground. The ground reality is far different from the assumption of policy makers. The beneficiaries must have *adhar* linked ration card. Most of beneficiaries are digital illiterate and in rural areas weak signals make the process even more daunting. Anil and Dreze (2021) has pointed to such challenges in absence of registered mobile number or email address and lack of information/communication for retrieval process. Many Fair Price Shops in Bihar charged Rs 20 as service charge from the beneficiaries.

The success of Sustainable Development Goals depends directly or indirectly on the success of SDG 3 (*Wellbeing*). SDGs *wellbeing* is based on the experiences of health target of MDGs. Health sector has been a major concern for Indian policy

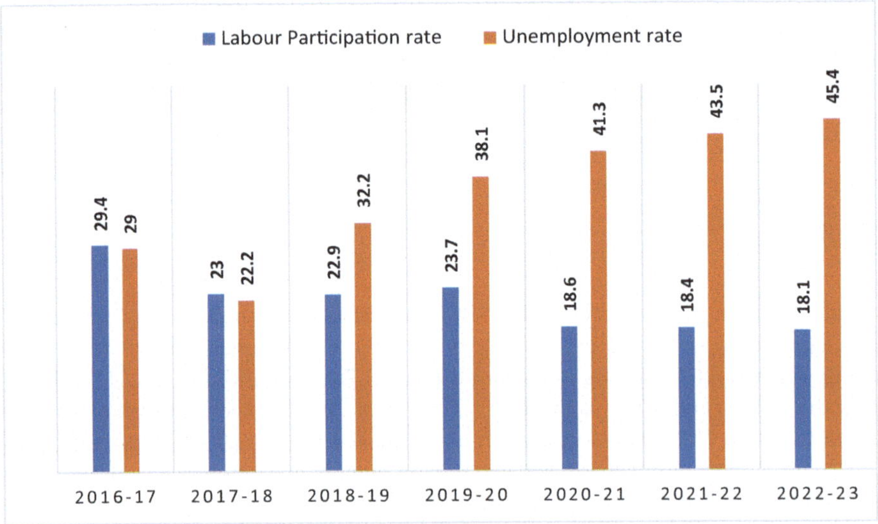

Fig. 2.11 Trends for Labor Force Participation and unemployment rate among Indian youth (15–24 years). *Source* Center of Monitoring Indian Economy (2023)

makers since the country has made considerable progress in achieving health target of MDGs. COVID19 adversely affected the public health infrastructure of India as the country struggled to provide basic health facility to infected persons and the burden of diseased population fell on an unprepared public health system. The country's weak health infrastructure was the result of underinvestment in the sector. In the absence of adequate public health sector, the people are forced to knock the door of exploitative private hospitals. Privately managed hospitals are unregulated, money making and extortionist in nature. The cruel pandemic forced Government to revamp the public health sector without which it would not be possible to achieve the health targets prescribed in SDG5. The annual budget (2021–22) showed a hike of 137% from the previous year in health budget.

Low participation of women in the labor force is a major challenge for sustainable development in India. India has achieved the target of women's participation in decision-making in the third tier of democracy, however, the success is not translated in the second and third tier. Female entrepreneurs face multiple challenges ranging from inadequate access to credit, capacity building, limited information, and networking linkages (NITI Aayog, 2022).

Although, the country has almost achieved the target of clean drinking water through improved sources of water, however, the quality of water is unhealthy and has a chronic impact on human health.

MSME sector was in distress before the impact of COVID19. This sector was under transition due to two policy changes, first demonetization and second implementation of GST. Most of the micro enterprises in India are too small to qualify the criteria of micro enterprises to escape from the ambit of GST norms. In normal times it worked in their favor because they were not maintaining accounts, not paying taxes which make their product cheaper and they earn good money, but in the time of crisis they collapsed because they do not fall under the guideline of MSME to secure the Government help. MSMEs and Non-Banking Financial Institutions and other commercial vehicle operators are under economic shock of the past years (Adian et al., 2020). NBFCs are facing new challenges due to contraction of economy. COVID19 adversely affected NBFCs as lockdown hiked the risk of liquidity and degradation of asset quality.

At the time when the country faced the crisis of the pandemic, the relevance of job security scheme MGNREGA became more evident. MGNREGA has its own problems in implementation. The biggest challenge is 'ghost laborer' as evident in the stories of fake job card, fake muster roll and the fake work site from time to time. Another problem is lack of financial literacy. Marginalized people hesitate to use technology due to fear of fraud which restrict them from using modern technology or app based financial system. As shown in Fig. 2.12, in the last four years, the scheme crossed the 50 days average work only once, the average number job days in financial year 2022–23 reduced the level of 2015–16 (Business Today, 2019; Mukherjee & Dhasmana, 2020; Lok Sabha, 2023). Further it shows the uneven distribution of job among the beneficiaries; some house hold gets more than average while some are below average.

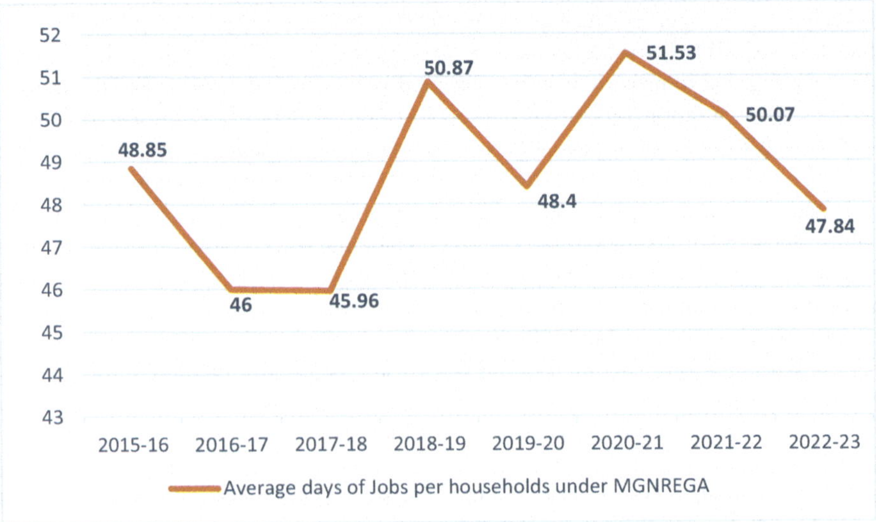

Fig. 2.12 Average job days under MGNREGA, Financial Year 2015–16 to 2022–23. *Source* Business Today (2019), Mukherjee and Dhasmana (2020), Lok Sabha (2023)

It was assumed that the worker is renumerated within 15 days of work. In case of any delay, they should be compensated. Narayanan et al. (2019) surveyed 90 lakh transactions across 10 states during 2016–17 and their finding suggest that only 21% of payment were made on time and that the Centre itself consume 50 days for electronic transfer. Interestingly, the Centre had acknowledged the delay but never paid the penalty for delay. In 17 states MGNREGA wages were lower than the lower agriculture wages. These challenges are sufficient to weaken the ACT made to solve the bread-and-butter issue of the poor living in rural India.

Conclusion

In the implementation of the first three SDGs, India shows progress in poverty reduction, significant improvement in maternal and child health, universalize the education system through NEP 2020, passed the women reservation bill to increase the proportion of women in national Parliament and State Assemblies. Access to 'improved drinking' has shown progress but there is still a great need to improve the access of clean piped water, since the improved source of water are not free from chemical contamination and not safe and healthy for drinking purpose. Rising inequality remains a major concern for policy makers. The COVID-19 pandemic has further adversely affected the progress made.

Provisions of stimulus package are for immediate relief of economy at the same time it makes various sector capable of creating more jobs. Government has tried to provide various relief measures and liberalized the restriction which provide level playing field for private investors. Starting from Pradhan Mantri Garib Kalyan Yojana to the restructuring of Medium Small and Micro Enterprises, relief to Non -Banking Financial Institutions, raising the FDI cap in strategic sector such atomic and defense, Government has tried to build trust to attract private investors which translates into savings resulting in expansion of economy.

However, with all the advancement in pre-COVID years and immediate stimulus relief during crisis, India is still home to the largest poor population in the world. Evidence suggests that in the long run, increasing employment rate had by far the biggest impact in poverty reduction. In order to reduce incidence of poverty, India has to create more jobs. Decent income in pocket provides opportunity in invest more in health and education and ultimately take the country on the path of achieving Sustainable Development Goals.

Appendix 2.1: Sex Ratio at Birth in Indian States

	NFHS 4 (2015–16)	NFHS 5 (2019–20)
Andaman & Nicobar	859	914
Andhra Pradesh	914	934
Arunachal Pradesh	926	979
Assam	929	964
Bihar	934	908
Chandigarh	981	938
Chhattisgarh	977	960
DNH & DD	983	817
Goa	966	838
Gujarat	906	955
Haryana	836	893
Himachal Pradesh	937	875
Jammu & Kashmir	923	976
Jharkhand	919	899
Karnataka	910	978
Kerala	1047	951
Ladak	823	1125
Lakshadweep	905	1056
Madhya Pradesh	927	956
Maharashtra	924	913
Manipur	962	967

(continued)

(continued)

	NFHS 4 (2015–16)	NFHS 5 (2019–20)
Meghalaya	1009	989
Mizoram	949	969
Nagaland	953	945
NCT of Delhi	812	923
Odisha	932	894
Puducherry	843	959
Punjab	860	904
Rajasthan	887	891
Sikkim	809	969
Tamil Nadu	954	878
Telangana	872	894
Tripura	969	1028
Uttar Pradesh	903	941
Uttarakhand	888	984
West Bengal	960	973

References

Abraham, R., Basole, A., & Kesar, S. (2021). *Tracking employment trajectories during the Covid-19 Pandemic: Evidence from Indian panel data*. Azim Premji University.

Adhikari, A., & Bhatia, K. (2010). NREGA wage payments: Can we bank on the banks? *Economic and Political Weekly, 45*(1), 30–37. https://www.jstor.org/stable/25663966.

Adian, I., Doumbia, D., Gregory, N., Ragoussis, A., Reddy, A., & Timmis, J. (2020). *Small and medium enterprises in the pandemic impact, responses and the role of development finance*. World Bank. https://documents1.worldbank.org/curated/en/729451600968236270/pdf/Small-and-Medium-Enterprises-in-the-Pandemic-Impact-Responses-and-the-Role-of-Development-Finance.pdf.

Afridi, F., Mukhopadhyay, A., & Sahoo, S. (2016). Female labor force participation and child education in India: Evidence from the National Rural Employment Guarantee Scheme. *IZA Journal of Labor & Development*,https://doi.org/10.1186/s40175-016-0053-y

Ahankari, A. S., Myles, P., Tata, L. J., & Fogarty, A. W. (2015). Banning of fetal sex determination and changes in sex ratio in India. *The Lancet Global Health*, Available athttps://doi.org/10.1016/S2214-109X(15)00053-4

Anand, S. (2023). Centre unveils plans to add 250GW renewable energy capacity in next five years. *mint*, pp. Availabe at: https://www.livemint.com/industry/energy/indian-government-plans-to-add-250-gw-of-renewable-energy-capacity-in-next-5-years-to-achieve-500-gw-target-by-2030-11680685655723.html.

Ananian, S., & Dellaferrera, G. (2024). *Employment and wage disparities between rural and urban areas*. https://www.ilo.org/wcmsp5/groups/public/---dgreports/---inst/documents/publication/wcms_911918.pdf: International Labour Organization.

Anil, V., & Dreze, J. (2021). Without Aadhaar, without identity. *The Indian Express*, https://indianexpress.com/article/opinion/columns/flaw-in-aadhaar-architecture-uidai-card-enrolment-7389133/.

Ayyagari, M., Demirguc-Kunt, A., & Maksimovic, V. (2014). Who creates jobs in developing countries? *Small Business Economics, 43*(1), 75–99.

Azam, M. (2012). *The impact of Indian job guarantee scheme on labor market outcomes: Evidence from a natural experiment. IZA DP No. 6548.* Institute for the Study of Labor. https://docs.iza.org/dp6548.pdf.

Azim Premji University & CII. (2021). *Reviving employment and livelihoods in India: Covid-19 and after.* Bengaluru, Azim Premji University & Confederation of Indian Industry (CII). https://cse.azimpremjiuniversity.edu.in/wp-content/uploads/2021/03/CII-Report-Employment-Livelihoods-1.pdf.

Azim Premji University. (2021). *State of Working India 2021: One year of Covid-19.* Bengaluru. Available at: https://cse.azimpremjiuniversity.edu.in/state-of-working-india : Centre for Sustainable Employment, Azim Premji University.

Azim Premji University. (2022). *Employment Guarantee during Covid-19: Role of MGNREGA in the year after the 2020 lockdown.* Centre for Sustainable Employment, Azim Premji University and NREGA Consortium. https://cse.azimpremjiuniversity.edu.in/wp-content/uploads/2022/10/MGNREGA_Covid_Survey_Report_Final.pdf.

Balakrishnan, P. (2022). *India's economy from nehru to modi: A brief history.* Permanent Black.

Bangladesh Bureau of Statistics. (2021). *Statistical Yearbook.* Available at: https://www.bbs.gov.bd/site/page/29855dc1-f2b4-4dc0-9073-f692361112da/Statistical-Yearbook: Bangladesh Bureau of Statistics, Government of the People's Republic of Bangladesh.

Behera, M., Mishra, S., Mohapatra, N., & Behera, A. R. (2020). COVID-19 Pandemic and micro, small and medium enterprises (MSMEs): Policy response for revival. *SEDME (Small Enterprises Development, Management & Extension Journal), 47*(3), 213–228. https://doi.org/10.1177/09708464211037485.

Bloom, D., & Canning, D. (2003). Health as human capital and its impact on economic performance. *The Geneva Papers on Risk and Insurance—Issues and Practice, 28*, 304–315.

Boben, B., & Jain, R. (2023). Female footprint in India's workforce. *Reuters*.

Business Today. (2019). Forget 150 days employment under MGNREGA! Numbers on ground not even half of existing 100 workdays. *Business Today*, pp. https://www.businesstoday.in/latest/economy-politics/story/congress-manifesto-150-workdays-forget-150-days-employment-under-mgnrega-numbers-on-ground-not-even-half-of-existing-100-workdays-184993-2019-04-03.

Chacaltana, J., Bonnet, F., & Garcia, J. M. (2022). *Growth, economic structure and informality.* ILO Working Paper 69. https://www.ilo.org/wcmsp5/groups/public/---ed_emp/documents/genericdocument/wcms_849574.pdf.

Chafe, Z., & Chowdhury, S. (2021). A deadly double dose for India's poor. *Nature Sustainability, 4*, 835–836.

Bharti, N. K., Chancel, L., Piketty, T., & Somanchi, A. (2024). *Income and Wealth Inequality in India, 1922–2023: The Rise of the Billionaire Raj.* World Inequality Lab. https://wid.world/wp-content/uploads/2024/03/WorldInequalityLab_WP2024_09_Income-and-Wealth-Inequality-in-India-1922-2023_Final.pdf.

Chancel, L., Piketty, T., Saez, E., & Zucman, G. (2022). *World inequlity report 2022.* World Inequlity Lab.

Chandrashekhar, V. (2020). 1.3 billion people. A 21-day lockdown. Can India curb the coronavirus? *Science*, Available at: https://www.science.org/content/article/13-billion-people-21-day-lockdown-can-india-curb-coronavirus.

Chapman, A. R., Brunelli, L., Foreman, L., & Kaemp, J. (2023). Promoting children's rights to health and well-being in the United States. *Lancet, 25*, 100577, Available at: https://doi.org/10.1016/j.lana.2023.100577.

Chattaraj, S. (2016). Organizing the Unorganized: Union Membership and Earnings in India's Informal Economy. *BSG Working Paper Series*, Available at: https://www.bsg.ox.ac.uk/sites/default/files/2018-05/BSG-WP015.pdf.

Clark, C. (1940). *The conditions of economic progress*. Macmillan.

Dang, H.-A.H., & Lanjouw, P. (2023). India: Inequality Trends and dynamics: The Bird's-eye and the granular perspectives'. In C. Gradín, M. Leibbrandt, & F. Tarp (Eds.), *Inequality in the Developing World* (pp. 157–179). Oxford Academic.

Dang, H.-A., Lanjouw, P., & Vrijburg, E. (2021). Poverty in India in the face of Covid-19: Diagnosis and prospects. *Review of Development Economics., 25*, 1816–1837.

Das, S., & Mocan, N. (2012). Analyzing the impact of the world's largest public works project on crime. *Nber working paper series. National Bureau of Economic Research*, 1–75.

Dasaraju, H., & Tambunan, T. T. (2023). *Role of micro, small and medium enterprises in achieving SDGs: Perspectives from emerging economies*. Palgrave Macmillan Singapore.

De Vries, G., Erumban, A., Timmer, M., Voskoboynikov, I., & Wu, H. (2012). Deconstructing the BRICs: Structural transformation and aggregate productivity growth. *Journal of Comparative Economics*, 211–227.

De, P. K., & Nagaraj, P. (2014). Productivity and firm size in India. *Small Business Economics., 42*, 891–907. https://doi.org/10.1007/s11187-013-9504-x

Deininger, K., & Liu, Y. (2013). *Welfare and poverty impacts of India's national rural employment guarantee scheme. Policy Research Working Paper, 6543*. Washinton DC: World Bank. https://documents1.worldbank.org/curated/en/832841468041451142/pdf/WPS6543.pdf.

Deininger, K., & Liu, Y. (2013). *Welfare and poverty impacts of India's national rural employment guarantee scheme*: *Evidence from Andhra Pradesh*. World Bank.

Delfmann, H., & Koster, S. (2016). The effect of new business creation on employment growth in regions facing population decline. *The Annals of Regional Science., 56*, 33–54. https://doi.org/10.1007/s00168-015-0738-1

Deshpande, A., & Kabeer, N. (2024). Norms that matter: Exploring the distribution of women's work between income generation, expenditure-saving and unpaid domestic responsibilities in India. *World Development Volume 174*, Available at: https://doi.org/10.1016/j.worlddev.2023.106435.

Dhamija, D. (2023). What is the primary source of drinking water for Indian households? *Center for Economic Data Analysis. Ashoka University*, pp. Available at: https://ceda.ashoka.edu.in/what-is-the-primary-source-of-drinking-water-for-indian-households/.

Dreze, J., Dey, N., & Khera, R. (2008). Corruption in NREGA: myths and reality. *The Hindu*, pp. https://www.thehindu.com/todays-paper/tp-opinion/Corruption-in-NREGA-myths-and-reality/article15148488.ece.

FAO, IFAD, UNICEF, WFP & WHO. (2023). *The State of Food Security and Nutrition in the World 2023. Urbanization, agrifood systems, transformation and healthy diets across the rural-urban contnuum*. FAO.

Fisher, A. (1935). *The clash of progress and security*. Macmillan.

Fomba, B. K., Talla, D. N., & Ningaye, P. (2023). Institutional quality and education quality in developing countries: Effects and transmission channels. *Journal of Knowledge Economy, 14*(1), 86–115.

Fontaine, F., Turjeman, S., Callens, K., & Koren, O. (2023). The intersection of undernutrition, microbiome, and child development in the first years of life. *Nature Communication, 14, Article No: 3554*, Availbel at: https://doi.org/10.1038/s41467-023-39285-9.

Garcia-Escribano, M., Mogues, T., Moszoro, M., & Soto, M. (2022). The spending challenge of achieving the SDGs in South Asia: Lessons from India. In M. R. Salgado & R. Anand (Eds.), *South Asia's Path to Resilient Growth* (pp. 29–61). International Monetary Fund.

Gates, M. F., & Gates, B. (2023). *Imagine a World: Where innovations could save the lives of 2 million more mothers and babies*. Bill & Melinda Gates Foundation.

Glaser, G. (2012). Base sustainable development goals on science. *Nature, 491, 35*, Available at: https://doi.org/10.1038/491035a.

Grossman, M. (1972). On the concept of health capital and the demand for health. *Journal of Political Economy., 80*(2), 223–255.

Gupta, R., Singhal, G. L., Goyal, A., Kaur, G., Purohit, N., & Prinja, S. (2021). Combatting the imbalance of sex ratio at birth: Medium-term impact of India's National Programme of Beti Bachao Beti Padhao in the Haryana State of India. *Health Policy and Planning., 36*(10), 1499–1507.

Hale, T., Angrist, N., Goldszmidt, R., Kira, B., Petherick , A., Phillips, T., …Tatlow, H. (2021). *A global panel database of pandemic policies (Oxford COVID-19 Government Response Tracker).* Available athttps://doi.org/10.1038/s41562-021-01079-8:NatureHumanBehaviour

Hammer, A., Keles, J. Y., & Olsen, W. (2022). Working lives in india: current insights and future directions. *Work, Employment and Society, 36*(6), 1139–1154. Available at: https://doi.org/10.1177/09500170221083511.

Hannum, E., & Buchmann, C. (2005). Global educational expansion and socio-economic development: An assessment of findings from the social sciences. *World Development., 33*(3), 333–354.

Heath, R., & Jayachandran, S. (2016). The causes and consequences of increased female education and labor force participation in developing countries. NBER Working Paper No. 22766. *National Bureau of Economic Research,* Available at: https://www.nber.org/papers/w22766.

Holmes, R., Morgan, J., & Hagen-Zanker, J. (2011). *Progress in providing employment for the poor: The national public works programme in India.* Overseas Development Institute.

Huang, S., Zhou, T., Xu, C., & Zheng, J. (2022). Does public health influence economic performance? Investigating the role of governance and greener energies for the case of China. *Front Public Health.* https://doi.org/10.3389/fpubh.2022.864736. PMID: 35425739; PMCID: PMC9001902.

Ianchovichina, E. (2012). What Is Inclusive Growth? In R. ArezkiM, C. A. Pattillo, M. G. Quintyn, & M. Zhu (Eds.), *Commodity price volatility and inclusive growth in low-income countries* (pp. 147–160). International Monetary Fund.

IFC. (2022). *The state of inequality report.* Institute for Competitiveness.

IIPS. (2020). *National family health survey, India 2019–2021.* Indian Institute of Population Studies.

IIPS. (2020). *NFHS-5.* International Institute of Population Studies.

ILO. (2002). *Decent work and informal economy.* Available at: /www.ilo.org/public/english/standards/relm/ilc/ilc90/pdf/rep-vi.pdf: International Labor Organization.

ILO. (2022). *ILOSTAT.* Available at: https://ilostat.ilo.org/data/#: International Labor Organization.

ILO. (2022). *The gender pay gap, hard truths and actions needed.* International Labor Organization.

ILO. (2023). *Spotlight on work statistics n°12.* International Labor Organization.

ILO. (2023). *Women and men in the informal economy: A statistical update.* International Labour Office.

ILO. (2023). *World employment and social outlook trends 2023.* International Labour Organization.

ILO. (Acessed on 12 December 2023). *Decent work.* Available at: //www.ilo.org/global/topics/decent-work/lang--en/index.htm: International Labour Office.

IMF. (2020). *Annual report of the executive board.* International Monetary Fund.

IMF. (2021). *Annual report of the executive board.* International Monetary Fund.

India Today. (2017). India might be hungry, but 50 per cent of our milk, fruits & veggies go to waste each year. *India Today,* pp. https://www.indiatoday.in/lifestyle/wellness/story/india-milk-fruits-vegetables-waste-perishable-products-assocham-lifest-1070558-2017-10-25.

Indrawati, S. M. (2015). *What you need to know about energy and poverty.* Available at: https://blogs.worldbank.org/voices/what-you-need-know-about-energy-and-poverty: World Bank.

IPU. (2023). *Women in parliament: Data on share of women in the single/lower chamber of parliament as of 1 January 2023.* Available at: https://data.ipu.org/women-ranking?month=12&year=2022: Inter-Parliamentary Union.

Iversen, T., & Rosenbluth, F. (2008). Work and power. The connection between female laborforce participationand female political representation. *The Annual Review of Political Science,* 479–495.

Jacob, N. (2013). *The impact of NREGA on rural-urban migration: Field survey of Villupuram District, Tamil Nadu. Report No. 137.* Centre for Public Policy Research. https://www.andhrauniversity.edu.in/img/pdf/research/137.pdf.

Jahan, S., & Mahmud, A. S. (2015). What Is Capitalism? *Finance and Development*, p. https://www.imf.org/external/pubs/ft/fandd/2015/06/basics.htm.

Kaldor, N. (1967). *Strategic factors in economic development.* School of Industrial and Labor Relations, Cornell University.

Kant, A. (2021). Message. In N. Aayog, *SDG India: Index and Dashboard 2020–21* (p. 4). NITI Aayog.

Kanungo, S., Chatterjee, P., Saha, J., Pan, T., Chakrabarty, N. D., & Dutta, S. (2021). *The Journal of Infectious Diseases, 224*(5), 15, S573–S583.

Kesar, S. (2023). Economic transition, dualism and informality in India: Nature and patterns of household-level transitions. *Review of Economic Development*, 2438–2469.

Khan, M. R., & Ara, F. (2006). Women, participation and empowerment in local government: Bangladesh Union Parishad perspective. *Asian Affairs, 29*(1), 73–92.

Khanna, M., Sarmiento-Saher, S., Scutaru, A., & Singh, S. (2021). *SMALL BUSINESS, BIG GROWTH How investing in SMEs creates jobs.* International Finance Corporation. https://www.ifc.org/content/dam/ifc/doc/mgrt/ifc-sme-report-2021-fa-digital.pdf.

Khera, R. (2008). Empowerment guarantee act. *Economic & Political Weekly, 43*(35), 8–10. https://www.jstor.org/stable/40278713.

Khera, R., & Nayak, N. (2009). Women workers and perceptions of the national rural employment guarantee act. *Economic and Political Weekly, 44*(43), 49–57. https://www.jstor.org/stable/25663707.

Kochhar, R. (2021). In the pandemic, India's middle class shrinks and poverty spreads while China sees smaller changes. *Pew Research Centre*, pp. https://www.pewresearch.org/short-reads/2021/03/18/in-the-pandemic-indias-middle-class-shrinks-and-poverty-spreads-while-china-sees-smaller-changes/.

Kumar, N., Darshan, Unnikrishnan, B., Kanchan, T., Thapar, R., Mithra, P., …Holla, R. (2014). Awareness and attitudes regarding prenatal sex determination, pre-conception and pre-natal diagnostic techniques act (PCPNDTA) among pregnant women in Southern India. *JC09–JC11 8(10)*, Available at: https://doi.org/10.7860/JCDR/2014/9789.5033.

Kuznets, S. (1966). *Modern economic growth: rate, structure and spread*. Yale University Press.

La Porta, R., & Shleifer, A. (2014a). Informality and development. *Journal of Economic Perspectives, 28*(3), 109–126.

Lewis, W. A. (1954). Economic development with unlimited supplies of labour. *The Manchester School*, 139–191.

Lok Sabha. (2021). *Unstarred question No. 2589.* Lok Sabha, Government of India. https://sansad.in/getFile/loksabhaquestions/annex/177/AU2589.pdf?source=pqals.

Lok Sabha. (2023). *lok sabha unstarred question no. 1945 answered on 01/08/2023: average days of employment under mgnregs.* Lok Sabha, Government of India. https://sansad.in/getFile/loksabhaquestions/annex/1712/AU1945.pdf?source=pqals.

Mankiw, N. G. (2008). What Would Keynes Have Done? *The New York Times*, p. https://www.nytimes.com/2008/11/30/business/economy/30view.html.

McKinsey. (2015). *the power of parity: advancing women's equality in India.* McKinsey Global Institute (MGI).

McMillan, M., Rodrik, D., & Sepulveda, C. (2017). *Structural change, fundamentals, and growth: A framework and case studies.* The World Bank.

Ministry of Finance. (2022). *Economic survey economic survey 2021–22.* Available at: https://www.pib.gov.in/PressReleseDetail.aspx?PRID=1894901: Ministry of Finance. Government of India.

Mint. (2023). Happy Children's Day 2023! Wishes and messages to share with your lovely kids on 'Bal Diwas'. *mint*, pp. Available at: https://www.livemint.com/news/india/happy-childrens-day-2023-wishes-and-messages-to-share-with-your-lovely-kids-on-bal-diwas-11699924971355.html.

Mint. (2023). India will manage 8% growth by 2030 if more women join workforce, says report. *Mint*, pp. https://www.livemint.com/economy/india-will-manage-8-pc-gdp-growth-by-2030-if-more-women-join-workforce-says-report-11698328560823.html.

Misra, K. (2019). Does historical land inequality attenuate the positive impact of India's employment guarantee program? *World Development, 124*,https://doi.org/10.1016/j.worlddev.2019.104662

Mitra, P. P. (2022). Active labor market policies in developing Asia: Case study of India. In L. A. Hondo (eds.), *Strengthening Active Labor Market Policies to Drive an Inclusive Recovery in Asia* (pp. 20–36). Asian Development Bank Institute.

MMSME. (2022). *Annual report 2022–23.* Ministry of Micro, Small, and Medium Enterprises, Government of India.

MoSPI. (2017). *Millennium development goals—final country report of India.* Ministry of Statistics and Programme Implementation.

MoSPI. (2023). *Periodic labour force survey (PLFS) annual report [JULY 2021—JUNE 2022].* Ministry of Statistics and Programme Implementation. Government of India.

MSME, Government of India. (2022). *Annual Report 2021–22.* Government of India. https://msme.gov.in/sites/default/files/MSMEENGLISHANNUALREPORT2021-22.pdf.

MSME, Government of India. (2021). *Annual Report 2020–21.* Ministry of Micro, Small and Medium Enterprises. Government of India. https://msme.gov.in/sites/default/files/MSME-ANNUAL-REPORT-ENGLISH%202020-21.pdf.

Mukherjee, S., & Dhasmana, I. (2020). Demand v reality: A sorry state of affairs under MGNREGA in a Covid-19 year. *Business Standard*, pp. https://www.business-standard.com/article/economy-policy/demand-v-reality-a-sorry-state-of-affairs-under-mgnrega-in-a-covid-.

Murarkar, S., Gothankar, J., Doke, P., Pore, P., Lalwani, S., Dhumale, G., ... Malshe, N. (2020). Prevalence and determinants of undernutrition among under-five children residing in urban slums and rural area, Maharashtra, India: a community-based cross-sectional study. *BMC Public Health, 20* No 1559, Available at: https://doi.org/10.1186/s12889-020-09642-0.

Napier-Raman, S., Rattani, A., Qaiyum, Y., Bose, V., Seth, R., & Raman, S. (2021). Impact of COVID-19 on the lives of vulnerable young people in New Delhi, India: A mixed method study. *BMJ Paediatrics Open, 5*(1), Available at: https://doi.org/10.1136/bmjpo-2021-001171.

Narayanan, R., Dhorajiwala, S., & Golani, R. (2019). Analysis of Payment Delays and Delay Compensation in MGNREGA: Findings Across Ten States for Financial Year 2016–2017. *Indian Journal of Labour Economics. 62*(1), 113–133.

NARSS. (2022). *National annual rural sanitation survey (NARSS) ROUND-3 (2019–20).* Available at: https://jalshakti-ddws.gov.in/sites/default/files/NARSS_Round_3_2019_20_Report.pdf: Department of Drinking Water and Sanitation (DDWS). Governmnet of India.

NAS. (2021). *National achievement survey(NAS)-2021 report card.* Available at: https://nas.gov.in/report-card/nas-2021: NCERT.

Nath, D. (2021). Migrant workers face debt, job loss and separation from families. *The Hindu*, pp. Avaiable at: https://www.thehindu.com/news/national/migrant-workers-face-debt-job-loss-and-separation-from-families/article34117669.ece.

Nath, K. J. (2003). Home hygiene and environmental sanitation: a country situation analysis for India. *nt Journal Environment Health Research*, S19–28.

Nayyar, D. (2019). *Resurgent Asia: Diversity in development.* Oxford University Press.

Nehru, J. L. (1947). *A tryst with destiny. Audio recording.* University of Cambridge.

NITI Aayog. (2017). *Voluntary national review report on implementation of sustainable development goals.* NITI Aayog.

NITI Aayog. (2018). *SDG India Index: Base Line Report 2018.* NITI Aayog.

NITI Aayog. (2018). *SDG India index: Baseline report 2018.* NITI Aayog.

NITI Aayog. (2021). *Sdg India Index 3.0.* NITI Aayog. https://sdgindiaindex.niti.gov.in/assets/Files/Target_Justification_2020-21.pdf.

NITI Aayog. (2021). *SDG India: Index & Dashboard 2020–21 Partnership in the decade of Action.* NITI Aayog.

NITI Aayog. (2022). *Decoding government support to women entreprenuers in India.* Niti Aayog.

NITI Aayog. (2023). *India national multidimensional poverty index: A progress review 2023.* NITI Aayog.

NITI Aayog. (2023). *National multidimensional poverty index: A progress review 2023.* NITI Aayog.

NSSO. (2023). *Multiple indicator survey in India: NSS 78th Round.* Avalable at: www.mospi.gov.in/sites/default/files/publication_reports/MultipleIndicatorSurveyinIndiaf.pdf: National Sample Survey Office. Government of India.

OECD. (2017). *The Pursuit of Gender Equality: An Uphill Battle.* OECD Publishing.https://doi.org/10.1787/9789264281318-en

Olsen, W., & Mehta, S. (2006). A Pluralist Account of Labour Participation in India. *GPRG-WPS-042. Global Poverty Research Group,* Available at: https://ora.ox.ac.uk/objects/uuid:a39712d8-83d9-4029-abf6-9910b4955b6f.

Pangarso, A., Sisilia, K., Setyorini, R., Peranginangin, Y., & Awirya, A. A. (2022). The long path to achieving green economy performance for micro small medium enterprise. *J Innov Entrep, 11,* 16. https://doi.org/10.1186/s13731-022-00209-4

Patwardhan, S., & Tasciotti, L. (2022). The effect of the Mahatma Gandhi National Rural Employment Guarantee Act on the size of outstanding debts in rural India. *Journal of Development Effectiveness, 15*(4), 353–372. https://doi.org/10.1080/19439342.2022.2103169.

Pennington, A., Maudsley, G., & Whitehead, M. (2023). The impacts of profound gender discrimination on the survival of girls and women in son-preference countries—A systematic review. *Health & Place, 79,* Available at: https://doi.org/10.1016/j.healthplace.2022.102942.

PIB. (2011). *Women Reservation in Panchayats.* Press Information Bureau. Government of India.

PIB. (2013). *Poverty Estimates for 2011-12.* Press Information Bureau of India.

PIB. (2021). *Mahatma Gandhi National Rural Employment Guarantee Act 2005 (MGNREGA).* Press Bureau of India.

PIB. (2023). *Health an integral component of social welfare, leitmotif for the government.* Press Information Bureau, Govermnet of India.

PIB. (2023). PM hails passage of The Constitution (One Hundred and Twenty-Eighth Amendment) Bill, 2023 in the Lok Sabha. Press Bureau of India.

Piketty, T., Yang, L., & Zucman, G. (2019). Capital accumulation, private property, and rising inequality in China, 1978–2015. *American Economic Review*, 2469–2496.

Porta, R. L., & Shleifer, A. (2014). Informality and Development. *Journal of Economic Perspectives, 28*(3), 109–126.

Prachitha, J., Dhume, A., & Subramanian, S. (2019). India in pursuit of millennium development Goals: Were the targets really feasible? *Journal of Developing Societies, 35*(1), 105–133.

Ram, K., & Yadav, S. (2021). The impact of COVID-19 on poverty estimates in India: A study across caste, class and religion. *Contemporary Voice of Dalit,* Available athttps://doi.org/10.1177/2455328X211051432

Rao, N. D., Kiesewetter, G., Min, J., Pachauri, S., & Wagner, F. (2021). Household contributions to and impacts from air pollution in India. *Nature Sustainability, 4,* 859–867.

Rathore, M. (2023). *COVID-19 impact on unemployment rate in India 2020–2022.* Hamburg: STATISTA. https://www.statista.com/statistics/1111487/coronavirus-impact-on-unemployment-rate/.

Ravi, S. (2014). Workfare as an effective way to fight poverty. *Brookings*, pp. Available at: https://www.brookings.edu/articles/workfare-as-an-effective-way-to-fight-poverty-the-mnrega/.

Ravi, S., & Engler, M. (2015). Workfare as an effective way to fight poverty: The case of India's NREGS. *World Development, 67*, 57–71. https://doi.org/10.1016/j.worlddev.2014.09.029

Ray, S., & Pullabhotla, H. K. (2023). The changing impact of rural electrification on Indian agriculture. *Nature Communications*, Available athttps://doi.org/10.1038/s41467-023-42533-7

REC. (2023). http://www.saubhagya.gov.in/dashboard. Available at: http://www.saubhagya.gov.in/dashboard: REC Limited.

Saini, M., Sengupta, E., Singh, M., Singh, H., & Singh, J. (2023). Sustainable development goal for quality education (SDG 4): A study on SDG 4 to extract the pattern of association among the

indicators of SDG 4 employing a genetic algorithm. *Education and Information Technologies*, 2031–2069.
Sen, A. (1990). More than a 100 Million Women Are Missing. *The New York Review of Books*, pp. https://www.nybooks.com/articles/1990/12/20/more-than-100-million-women-are-missing/.
Sen, A. (1993). Capability and well-being. In M. Nussbaum & A. Sen (Eds.), *The Quality of Life* (pp. 30–53). Oxford University Press.
Sen, A. (1998). *Public Action to Remedy Hunger.* Forum for Free Enterprise.
Sen, A. (1999). *Development as freedom.* Oxford University Press.
Sen, A. (2002). Why health equity? *Health Economics*, 659–666.
Sharma, A., Saluja, M. R., & Sarma, A. (2016). Macroeconomic Impact of Social Protection Programmes in India. *Economic and Political Weekly, 51*(24), 121–126. https://www.jstor.org/stable/44003611.
Singh, U. K. (2022). *Human development in Bihar, India: Experiences of millennium development goals.* Springer Nature.
Singh, U. K. (2022). Voluntary party quota: An alternative route to achieve women's representation. *The Quint*, pp. Available at: https://www.thequint.com/gender/women-reservation-in-politics-alternate-route-quota-in-parties.
Slater, J. (2020, August 31). India's economy contracts by nearly 24 percent, its sharpest drop on record. *The Washington Post*, pp. https://www.washingtonpost.com/world/asia_pacific/indias-economy-contracts-by-nearly-24-percent-amid-pandemic/2020/08/31/92318fbe-eb70-11ea-bd08-1b10132b458f_story.html.
Somayya, N. (2023). *Youth unemployment shockingly high.* Available on: https://www.cmie.com/kommon/bin/sr.php?kall=warticle&dt=20230926184023&msec=816: Center of Monitoring Indian Economy.
Soni, P. (2021). Effects of COVID-19 lockdown phases in India: an atmospheric perspective. *Environ Dev Sustain*, 12044–12055.
Srivastava, N., & Srivastava, R. (2010). Women, work, and employment outcomes in rural India. *Economic and Political Weekly., 45*(28), 49–63.
Srivastava, R. (2020). Growing Precarity, circular migration, and the lockdown in India. *The Indian Journal of Labour Economics, 63*(Suppl 1), 79–86.
Stiglitz, J. E. (2021). From manufacturing-led export growth to a twenty-first century inclusive growth strategy: Explaining the demise of a successful growth model and what to do about it. In C. Gradín, M. Leibbrandt, & F. Tarp, *Inequality in the Developing World* (pp. 284–318). Oxford Academic. Available at: https://doi.org/10.1093/oso/9780198863960.003.0012.
Syrquin, M. (1988). Patterns of structural change. In H. C. Srinivasan, *Handbook of Development Economics, 1* (pp. 203–273). Elsevier Science Publishers.
The Constitution of India. (1950). *The Constitution of India.* Government of India.
The Economist. (2020). India's Economy has Suffered even more than most. *The Economist*, pp. Available at: https://www.economist.com/asia/2020/05/23/indias-economy-has-suffered-even-more-than-most.
The Telegraph. (2021). House praise for employment scheme Modi panned. *The Telegraph*, pp. Available at: https://www.telegraphindia.com/india/house-praise-for-employment-scheme-modi-panned/cid/1806621.
Tripathi, S., & Yenneti, K. (2020). Measurement of Multidimensional Poverty in India: A State-level Analysis. *Indian Journal of Human Development*, 257–274.
Turangi, S. (2021). MGNREGS Performance (2006–21): An Inter-State Analysis. *South Asia Research., 42*(2), 208–232.
UDISE+. (2019). *Unified district information system for education plus (UDISE+) 2018–19.* Ministry of Education. Government of India.
UDISE+. (2022). *Unified district information system for education plus (UDISE+) 2021–22.* Ministry of Education. Governmnet of India.
UN. (2015). *Resolution adopted by the general assembly on 25 September 2015.* United Nations.
UN. (2022). *Human development report 2022.* United Nations.

UN Women. (2023). *Local government country profile: Data on share of women in local government as of 1 January 2023.* Available at: Available at: https://localgov.unwomen.org: UN Women.
UNDP & OPHI. (2023). *Global multidimensional poverty index 2023.* United Nations Development Programmes and Oxford Poverty & Development Initiative.
UNDP. (2019). *Human Development Report 2019.* United Nations.
UNDP. (2022). *Profiling poverty beyond income shows where to have the biggest impact in a world in crisis.* United Nations Development Program.
UNESCO. (2020). *Global Education Monitoring Report: Gender Report 2020.* UNESCO.
UNFPA. (2020). *State of World Population 2020.* United Nations Populations Fund.
UNFPA. (2023). *Son preference.* Available on: https://www.unfpa.org/son-preference#summery146312. Acessed on: 12/12/2023: United Nations Population Fund.
UNFPA India. (2022). *The Sexual & Reproductive Health Status of Young People in India Challenges and opportunities for healthy outcomes.* Available at: https://india.unfpa.org/sites/default/files/pub-pdf/brief2_-_the_srh_status_of_young_people.pdf: United Nations Population Funds.
UNFPA India. (2024). Opinion: *The Key To Fast-Track India's Progress Is More Women In Local Bodies.* United Nation Population Fund. https://india.unfpa.org/en/news/opinion-key-fast-track-indias-progress-more-women-local-bodies
UNICEF. (2022). *Levels & trends in child mortality: Report 2022.* United Nations Children's Fund.
UNICEF. (2023). *Maternal mortality.* Available at: https://data.unicef.org/topic/maternal-health/maternal-mortality/: UNICEF.
UNICEF, WHO, & World Bank. (2023). *Levels and trends in child malnutrition: Key findings of the 2023 edition.* UNICEF.
United Nations. (2021). *Highlights 2020–2021.* New York: Published by the United Nations, Department of Economic and Social Affairs. https://cdn.un.org/desa/Highlight_Report_2020-2021.pdf?_gl=1*ooaqdr*_ga*NTg1ODQ4NzA2LjE2MzEzNTY2NTg.*_ga_TK9BQL5X7Z*MTcwOTE0MDczMy43OS4wLjE3MDkxNDA3MzMuMC4wLjA.
Verma, D. (2017). Preface. In Mo, *Millennium Development Goals—Final Country Report of India* (p. 5). Ministry of Statistics and Programme Implementation.
Vyas, M. (2021). *Unemployment rate in India. 15 million jobs lost in May 2021.* Center for Monitoring Indian Economy. https://www.cmie.com/kommon/bin/sr.php?kall=warticle&dt=20210601180645&msec=766.
Vyas, M. (2023). *Employment in India.* Centre for Monitoring Indian Economy Pvt. Ltd. https://www.cmie.com/kommon/bin/sr.php?kall=warticle&dt=20210601180645&msec=766. Accessed on: 15/12/2023.
Wheebox. (2023). *India Skills Report 2030.* New Delhi. Availabl at: do3n1uzkew47z.cloudfront.net/siteassets/pdf/ISR_Report_2023.pdf: Wheebox.
WHO. (2011). Preventing gender-biased sex selection: an interagency statement OHCHR, UNFPA, UNICEF, UN Women and WHO. World Health Organisation.
WHO. (2018). *Violence Against Women Prevalence Estimates, 2018.* World Health Organizations.
WHO. (2021). *Ambient (outdoor) air pollution.* World Health Organization.
WHO. (2021). *Malnutrition.* World Health Organization.
WHO. (2022). *Nutrition Landscape Information System (NLiS): Improved sanitation facilities and drinking-water sources.* Available at:https://www.who.int/data/nutrition/nlis/info/improved-sanitation-facilities-and-drinking-water-sources: World Health Organization.
WHO. (2023). *Child health.* Available at: https://www.who.int/health-topics/child-health#tab=tab_1. Accessed on 05/12/2023: World Health Organization.
WHO. (2024). *WHO COVID-19 dashboard.* World Health Organizations.
Wilmoth, J. R., Mizoguchi, N., Oestergaard, M. Z., Say, L., Mathers, C. D., Zureick-Brown, S., ... Chou, D. (2012). A new method for deriving global estimates of maternal mortality. *Stat Politics Policy,* Available athttps://doi.org/10.1515/2151-7509.1038
World Bank. (1993). *World Development Report 1993: Investing in Health.* The World Bank Group.
World Bank. (2016). *Government of India and world bank sign US$1.5 billion agreement to support India's Universal sanitation initiative.* The World Bank.

World Bank. (2019). *mall and Medium Enterprises (SMEs) Finance: Improving SMEs' access to finance and finding innovative solutions to unlock sources of capital.* World Bank. https://www.worldbank.org/en/topic/smefinance.

World Bank. (2019). *Small and Medium Enterprises (SMEs) Finance. mproving SMEs' access to finance and finding innovative solutions to unlock sources of capital.* World Bank. https://www.worldbank.org/en/topic/smefinance#:~:text=In%20emerging%20markets%2C%20most%20formal,emerging%20markets%20and%20developing%20countries.

World Bank. (2020). *Global Economic Prospects, June 2020.* World Bank.https://doi.org/10.1596/978-1-4648-1553-9

World Bank. (2021). *Informal Economy Database.* Available at: https://www.worldbank.org/en/research/brief/informal-economy-database: World Bank.

World Bank. (2022). *Report on evolution of Ayushman Bharat Pradhan Mantri Jan Arogya Yojana, India.* World Bank India.

Chapter 3
Sustainable Development Goals in South Africa: Progress and Challenges

Introduction

On 27 April 1994, South Africa broke free from the shackles of 300 years of colonial rule, 84 years of racist Union of South Africa, and 46 years of statutory apartheid and emerged as an independent state. And yet, even in the post-apartheid era, South Africa had to deal with the legacy of apartheid including the well-entrenched poverty and inequality that were bestowed by the apartheid regime.

A new constitutional democracy was set up with the commitment to redress historical injustice and discrimination (Nagar, 2023). After independence in 1994, South Africa reconfigured the existing policies to be more inclusive through constitutional provisions and entitlement (Cirillo & Tebaldi, 2016). The constitutional mandates include the inclusion of social, political, economic, cultural, and environmental rights that are enforceable by the courts of law.

The duty of the government was re-envisioned to protect the rights and improve the well-being of all South Africans so that every individual has the opportunity to realize their full potential and contribute to nation-building. The country adopted right-based approaches and with more public spending and infrastructure, has emerged as a country with a vast coverage of social safety net programs, unlike the low-income countries depending on international donor agencies. The Brooking report noted the reduction in the rate of poverty and inequality in South Africa resulting from social assistance programs (Bhorat & Cassim, 2014).

South Africa is the driving force of the African economy. According to the African Development Bank, South Africa is the most industrialized among the 54 countries of the African continent (AfDB, 2022). As per IMF (2023) estimation, South Africa is the second largest economy in Africa. The economic progress in the post-apartheid period has achieved remarkable progress in poverty reduction, improved access to basic services, health, education, and social protection, and opened avenues of economic opportunities. The progress in the social sector has managed to leverage

U. K. Singh, *Regional Dimensions of Human Development in India and South Africa*, Advances in Geographical and Environmental Sciences,
https://doi.org/10.1007/978-981-97-5846-3_3

some of the adverse effects of institutionalized racial segregation under the apartheid regime.

World Bank literature shows that South Africa has registered significant progress in poverty reduction with the advent of democracy (World Bank, 2023a). However, the country has experienced the reversal of poverty reduction between 2011 and 2015. Recent data shows that an estimated half of South Africans are chronically poor as per the upper-bound national poverty line (2015 base price) of ZAR 992 (World Bank, 2018a). The high concentration of poor and low proportion of people with high-income earners along with the small share of middle-income earners makes South Africa a highly income-polarized country. Rising unemployment and sluggish growth rates remain the major concerns of the sustainable development process.

Aforesaid feature of South Africa makes it something of an enigma; on one side, it represents one of the most unequal regions in the world (World Bank, 2022a, 2022b, 2022c, 2022d) with half of the population in abject poverty, and on the other side, the country demonstrate high per capita income of $6766 (current USD) (World Bank, 2023a, 2023b, 2023c, 2023d). The study of Fransis and Webster (2019) indicates poor chances to reversal of poverty rate in near future. South Africa is country with plentiful natural resources and low population density and developed infrastructure (Munslow & Fitzgerald, 1994).

Practice of Sustainable Development in South Africa

The practice of sustainable development is rooted in the tradition of South Africa. For South Africans, the ideal of '*Ubuntu*' (I am because you are) is the guiding light to maintain harmony between human beings and mother nature. Nelson Mandela, arguably South Africa's most influential and prominent leader, recognized *ubuntu* as an encouraging force to hold the idea of social cohesiveness. In a TV interview Mandela explained the meaning of Ubuntu with the example of a society that is based on sharing (what we have) and caring (help others in need). Mandela's vision of a just, equal, and sustainable world order inspired leaders across the world. His ideas are aligned with the SDGs and serves as guiding principles to make a difference in society and guide humanity on the path to realize a better world (UN, 2023a, 2023b, 2023c).

The commitment for Sustainable Development was reflected in the inauguration speech of Nelson Mandela, as the President of the Republic of South Africa. Widely acclaimed as a hero of humanity, Mandela declared "*In our routine activities as a normal South African we must yield a real South African reality that strengthen human belief in justice, reinforce confidence in the nobility of mankind to endure our hopes for glorious life for all South Africans*". Mandela urged the South Africans to pledge to liberate themselves and their countrymen from poverty, inequality, suffering, gender, and other forms of discrimination, and to adopt a commitment to achieve peace and justice. He envisioned the victory of South Africa as the shared victory of human dignity as well as justice and peace.

The objective of SDGs includes the restoration of people's dignity along with other development targets. It becomes more relevant in South Africa with its history of injustice, racial segregation, and deprivation of the mass of the population. The implementation of SDGs is a reminder of the promise of independence to make an inclusive and progressive South Africa for all South Africans. The country has achieved some success during MDGs but there is still a need to reenergize the resources to accelerate inclusive and sustainable economic growth to complete the task of freeing people from misery and making a truly just South Africa.

To continue the development journey in achieving the remaining goals, South Africa has adopted the global goals in the form of Sustainable Development Goals for the next 15 years. Speaking at the General Assembly conference (2014), the then South African president said, "*in spite of improvement we have scored thus far, report continue to show that South Africa is lagging behind in achievement of some of the MDGs. That's why we appreciate crafting of new net of post-2015 goals to finish the unfished task of MDG*" (UN, 2014). South Africa adopted this report as the base line for Sustainable Development Goals (SDGs). The report reflects success and challenges, lesson learned for policy makers to build coherence between the post-2015 goals and Sustainable Development Goals in a way to realize the country's national development goals, to reduce poverty, and hunger, and to ensure access to health facilities and curb inequality by 2030.

South Africa integrated the African Union agenda 2063, Southern African Development Community's Regional Indicative Strategic Development Programme and aligned its National Development Programme in line with the Sustainable Development Goals, to make the country free from deprivation and inequality. Goals are incorporated into Government policies and processed at all level of democracy.

The final report (2015) highlights the progress and shortcomings of the Millennium Development Goals. Table 3.1 shows that South Africa South Africa performed well on all goals and indicators but lagged in the targets of health-related goals.

South Africa registered substantial progress in the MDGs period (2000–2015). The country successfully reduced the proportion of poor (MDG1) to 7.4% (2014) from 14% (2006) and witnessed significant contraction in hunger incidences. It also attained considerable success in MDG2, access to primary education, registering 99.3% learner enrolled in primary education till 2013. It scored well on MDG3, women empowerment, and performed well in attainment of GER (Primary and Tertiary), GPI and literacy rate among women, but lagged behind in female representation in parliament, and proportion of women in non-farm work. Iimpressive progress was noted in MDG4 (Infant mortality) and MDG5 (Maternal Health) but fell short in achieving targets. It did well in attainment of MDG7 and MDG8, but investment in GDP and non-availability of data for endangered species raised concern for the policy makers.

The country's stride in education, health, access to clean water, electricity and sanitation were recognized. During the MDGs period, access to primary education expanded to over 9 million poor children who attended 'no-fee' schools. The reach of social security extended to 17.5 million in 208 from 3 million in 1994. South Africa also rolled the world's biggest anti-retroviral programme to 4.5 million people.

Table 3.1 MDGs success card, South Africa (2015)

MDG1 Reduce poverty by one third	2015 Target	Status
Share of population living below USD1.25(PPP)/day	8.5%	7.4% (2011)
Share of population below USD 2.00(PPP)/day	16.8%	20.8% (2011)
Poverty gap ratio {USD 1.25 (PPP)/day}	8.5%	1.9% (2011)
Poverty gap ratio {USD 2.00 (PPP)/day}	6.5%	6.5% (2011)
Poverty gap ratio {USD 2.50 (PPP)/day}	9%	10.3% (2011)
Proportion of poorest quintile is national consumption	5.8%	2.11% (2011)
Share of Households SAMPI	No target	8% (2011)
Intensity of SAMPI	No target	42.3% (2011)
SAMPI score	No target	0.03% (2011)
Growth rate (percentage)/per person employed	6%	−1.1% (2013)
Employment-population ratio	50–70	42.8% (2014)
Proportion of employed person living below USD1 (PPP)/per day	0	3.9% (2009)
Proportion of self-account and active participant family worker in total employment	5%	9.3% (2009)
Proportion of population reporting hunger	15%	12.9% (2011)
Incidence of underweight (under 5 year of age)	4.7%	8.3% (2008)
Incidence of stunting (under 5 year of age)	15%	23.9% (2015)
MDG2: ensure access of universal Primary Education		
Net enrolment ratio adjustment in primary education	100%	99.3% (2013)
Share of learner reached to last primary grade starting from 1 grade	100%	96% (2013)
Literacy rate (15–24) year of age	100%	94% (2013)
National senior certificate pass rate	75%	76% (2014)
Net enrolment ratio adjustment in tertiary education	20%	19.4% (2013)
Under Graduated pass	35.6%	28% (2013)

(continued)

Table 3.1 (continued)

MDG1 Reduce poverty by one third	2015 Target	Status
MDG3 Gender equality and women empowerment		
GER and GPI (Primary Level)	01:01	0.99:1 (2013)
GPI (Secondary)	01:01	1.03:1 (2013)
GPI (Tertiary)	01:01	1.41:1 (2013)
Literate female to literate male ratio (15–24 year of age)	01:01	1.05:1 (2013)
Female proportion in no-farm wage employment	50%	45% (2013)
Unemployed female to unemployed male ratio (15–64 year of age)	01:01	1.0:1 (2013)
Share of women representation in nation parliament	50	42
MDG4 Reduce infant mortality		
IMR (Under 5 year of age)	20	34.3 per 1000 (2013)
IMR	18	23.6 per 1000 (2013)
Percentage of children (1 year of age) immunized against measles	100	91.2 (2014)
MDG 5 Maternal Health		
Maternal Mortality ratio	38	141 per 100,000 (2013)
Percentage of deliveries attended by skilled health personnel	100%	91% (2013)
Institutional deliveries	96%	85.6% (2014)
Rate of contraceptive prevalence	100%	50.6% (2003)
Couple year protection rate	No target	52.7% (2014)
Antenatal care handling	100%	92.9percent (2014)
MDG6 reverse the spread of communicable diseases		
Prevalence of HIV (15–24 year of age)	8.7%	7.1% (2012/13)
Prevalence of HIV among pregnant women (15–24 year of age)	22.8%	21.7% (2012)
Use of Condom at last risk sex	75.9	58.4% (2012/13)
School attendance proportion of orphan to non-orphan (10–14 year of age)	01:01	1:1 (2012)
Malaria cases	<64,600	8851 incidences (2013)
Death rate (malaria incidence)	<2.0/100000	0.6 per 100,000 (2013)
TB incidence	<253/100000	860 incidences (2013)

(continued)

Table 3.1 (continued)

MDG1 Reduce poverty by one third	2015 Target	Status
Death rate (TB)	<147/100000	76 per 100,000 (2013)
Percentage pf TB incidences identified and treated under DOTS	100%	90% (2012/13)
MDG 7 Environment Sustainability		
Ozone depleting substance consumption: hydrochlorofluorocarbon	Reduce HCFC by 10%	284.8 ODP mton (2013)
Ozone depleting substance consumption: bromochloromethane	100% Reduction by 2002	0 ODP mton (2013)
Ozone depleting substance consumption: methyl bromide	Phase out the consumption of MeBr	140.5 ODP mton (2013)
Proportion of endangered species	Data not available	Data not available
Share of population using improved water source	88.3%	90.8% (2013)
Share of population using improved sanitation facility	74.7%	76.8% (2013)
MDG 8 Partnership for Development		
Proportion of investment in GDP	25%	20.1% (2013)
Consumer price index inflation	3–6%	5.7% (2013)
Telephone line per 100 population	Universal Access	7.1 per 100 (2013)
Cellular telephone per 100 population		145 per 100 (2013)

Source Department of Statistics, RSA (2016)

Investment levels increased which in turn translated into more job opportunities (UN, 2019). However, notwithstanding the significant progress in South Africa's developmental journey since the dawn of democracy, challenges of poverty, unemployment and inequality, continue to pose a major threat towards forming an egalitarian society. Agenda 2030, explicitly addresses inequality in its agenda call for '*no one will be left behind*'.

Sustainable Development Goals and Its Implementation in South Africa

South Africa has recognized the importance of achieving global goals in the regional context in the pursuit of human wellbeing and commitment to uplift humanity so that "no one will be left behind" (Ntshavheni, 2023). For South Africa, SDGs mean a lot more than an international promise (Ramaphosa, 2023). The SDGs provides an opportunity to aligned the regional, national and continental and international development priorities to achieve more inclusive outcomes.

Policy coherence is vital to address the multidimensional challenges of development and a key element in global development discourse. It becomes more relevant in designing policies in respect of SDGs implementation (Verschaeve et al., 2016). The successful attainment of goals is dependent on the convergence of all development objectives envisaged in national, continental and international agendas (Ansell et al., 2022; Doorgapersad, 2022). South Africa has implemented SDGs within the local and continental agendas such as Agenda 2063 floated by African Union, Regional Indicative Strategic Development (RISDP) of Southern African Development Community's (SADC), and its own National Development Plan (NDP) (UN, 2017; Royo et al., 2022).

Agenda 2063 is the blue print to transform Africa into a global powerhouse. Its prime objective includes sustainable and inclusive future of African continent (African Union, 2015). RISDP envisaged inclusive, competitive and peaceful society to all member countries of SADC (2020). South Africa's NDP is a long-term development strategy. Its vison encapsulates the people's concern regarding everyday challenges, and is designed to take South Africa on the path of sustainable development by prioritizing inclusive development in national planning. South Africa played a key role in drafting of Agenda 2063, and it is significantly aligned with South Africa's NDP.

Localization of goals necessitate participation of both sides, on supply side SDGs provide a blueprint for local development policies, and policy response of local and regional government to achieve SDGs (Ngubane & Pillay, 2023). National policies prioritized sustainable economic growth and inclusive development thorough partnership with multiple stakeholders of development including public sector, private sector and civil society and more importantly academia and research institutions (Haywood et al., 2018).

Implementation and accomplishment of SDGs depends on employing development strategies based on regular measurement and assessment of indicators (Liu et al., 2023). Quantitative measurement of SDGs progress has been provided by the global (Sachs et al., 2023), regional (Allen et al., 2017) national (Allen et al., 2020) and subnational level (Xu et al., 2017) by several public, private and international agencies. To report the measurement of SDGs in South Africa, a dedicated integrate indicators framework (IIF) has been developed to coordinate regional, national and global agendas (Maluleke, 2023a, 2023b).

The success of development policy objectives is dependent on the state's delivery capacity. So as for SDGs objectives, all the stakeholder must ensure convergence of SDGs to NDP. Department of Planning, Monitoring and Evaluation (DPME), has the mandate from the South African parliament to coordinate policy design and implementation, and is also accountable for monitoring performance.

Development Instrument Instituted by Government Since 1994

Mandela's legacy and his commitment for justice, peace, equality, and human dignity has been carried by his successors. The United Nation recognized his ideas as a nucleus of Sustainable Development. In the honour of his legacy of reconciliation, political and social transformation, the Nelson Rolihlahla Mandela Prize was established for those who dedicated their life in service of humankind (UN, 2015a, 2015b, 2015c).

Mandela and his successive Governments emphasized job creation and poverty reduction to meet income equality. These ideas were translated into various development policies to ensure sustainable economic growth and translate growth into development to provide an umbrella of social security nets with competent delivery system. These development frameworks include Reconstruction and Development Program (RDP) which was the first economic plan of post-apartheid era to embark on the task to build a new nation (RSA, 1994; Drewes & Aswegen, 2013). To realize the commitment of Mandela, the African National Congress (ANC) framed the Reconstruction and Development Programme (Corder, 1997). RDP was a people driven programme and its success depended on full participation of both the rich and poor. RDP was replaced in 1996 by a new set of reform 'Growth Employment and Redistribution' (GEAR) segregated into Urban Development Framework and Rural development Framework (1997) for inclusive development (DHS, 1997). In 2005, to halve the rate of unemployment, reduce the poverty by fifty percent, and curb the menace of unemployment and to promote black economic empowerment, the Government introduced Accelerated and Shared Growth Initiative for South Africa (ASGISA) (McGrath & Akoojee, 2007). In 2007, the Government's first intervention in Industrial policy was seen and the National Industrial Policy was approved in January 2007 (DTI, 2007). There was a strategic shift from orthodox industrial policy based on Laissez-fair to policy based on Washington Consensus (Zalk, 2014). In 2001, South African Government tried to provide a common platform to all Government departments through Integrated Sustainable Rural Development Strategies (ISRDS) which was launched to accelerate sustainable economic growth by integrating urban and rural economy (OIS, 2000). In 2010 the New Growth Plan was introduced for reduction of unemployment by increasing state investment in energy, transport, communication, water and housing. In 2009, the newly elected president Jacob Zuma introduced two new institutions: Department of Economic Planning and The National Planning Commission. The Department of Economic Planning rolled out the National Growth Plan to prioritize employment generation in all Government policies (Fine, 2012). In 2010 South Africa established its National Planning Commission (NPC) to conceptualize long-term development program. The Commission identified the key areas responsible impeding poverty reduction and expansion of the economy which included poor quality of education, soaring unemployment, overburdened health sector, unequal society, failed public service, uneven distribution of poverty, weakened infrastructure, and lack of technology. To deal with these challenges, the

Commission introduced the long-awaited National Development Programme, 2012 to eliminate poverty in all forms and to provide decent jobs to all and reduce inequality with the target year of 2030.

Being a member of the United Nations, South Africa is a signatory of the Global Agenda –2030. Building on the experiences of the Millennium Development Goals, South Africa is committed to achieving the Sustainable Development Goals by 2030 through its national development targets.

The following sections analyses the transition of South Africa from MDGs to SDGs. It explores the progress achieved towards SDGs during the first 3 years of implementation. Further, it explores the impact of COVID-19 on the progress made and attempts to propose a way forward. The discussion will more specifically focus on the SDG goals of no poverty, zero hunger, health, quality education, gender equality, water and sanitation, inequality, and decent jobs.

Advent of National Development Program in South Africa

The first task assigned to the NPC was to identify the challenges of developmental programme implemented in the country since 1994. NPC presented a critical assessment as Diagnostic Report (2011) which indicated that in the journey of development, South Africa registered upward trends in post-apartheid era by protecting people from any type of discrimination, providing constitutional and legal framework to ensure equal rights and provided basic necessities under a democratic setup (National Planning Commission, 2011). However, gaps were identified as constitutional provisions are not fully implemented and effective focus fades with the passage of time. Moreover, the apartheid legacy continued to retard equal opportunity and adversely affected the down trodden citizen of the country.

The Commission identified education and employment as the most persistent challenges facing South Africa. In the failure of these two areas, development programs became futile in generating jobs and raising employment (World Bank, 2021). Unequal public spending in education sector negatively reflected in dismantling the vision of free schooling to poor children. Studies have shown that countries can improve employment levels with better education infrastructure build on public finance. Job opportunities for working population increases with quality education which in turn leads to raising the purchasing power translating into more money in pocket and decreased dependency of dole, accumulate savings to reinvest in market. In that way the economy expands and citizens enjoy the benefits of political and social security. The NPC assessment report was considered as a constructive step in drawing the long vision of NDP in November 2011 and implementation in 2012. The NDP draft delivers a detailed examination of the structural factors that need to be addressed to realize the Vision 2030 as given in Table 3.2.

The second task for NPC was to develop a vision document (vision 2030) to meet the challenges identified. The then President Jacob Zuma, at the inaugural meeting of Planning Commission stated "*we are committed to building a state that will expand*

Table 3.2 The challenges of South Africa

Sr no	Challenges
(1)	Thin work force
(2)	Poor quality of education for black children
(3)	Inadequate infrastructures are not well maintained and poorly located
(4)	Spatial division hamper inclusive division
(5)	Economy is resource intensive
(6)	Inadequate health infrastructure failed in meeting the demands
(7)	Uneven distribution public services
(8)	High level of corruption
(9)	Persistent inequality

Source Diagnostic Report, National Planning Commission South Africa (2011)

Table 3.3 Objects of NDP

1	Providing that overreaching goals for what South Africa would like to achieve by 2030
2	Call consensus to overcome the obstacles to achieving goals
3	Framework for long term goals within NDP
4	Guideline to choose best from scarce resource

Source National Development Plan (2012)

economy, reduce poverty and improve standard of living. Planning Commission mandate is to help define South Africa we desired to achieve in 20 years from today" (National Planning Commission, 2011).

NDP: Vision 2030 was meant to translate the aspiration into action and implement domestically designed measures to facilitate all the key instruments of development. The National executive in 2013 rolled the first Medium Term Strategic Framework (MTSF) 2014–2019 which was the first 5five year kit for NDP. Vision 2030 restated the key obstacle in realizing faster growth under RDP, GEAR, ASGISA and the NGP among other things. Based on the experiences of instruments of development used by the Government of South Africa, NDP gives a clear roadmap for sustainable and inclusive development. NDP summaries its vision in four broad areas as given in Table 3.3.

Convergence of NDP and SDGs

As a signatory of the 70th General Assembly, United Nation resolution (25 September 2015), South Africa pledged to achieve the global goals in form of SDGs by 2030. Recognizing the attainment of SDGs as an obligation, the Government of South Africa aligned NDP provisions with SDGs. United Nation Development Programme

(UNDP) conducted a study to identify common area objectives between NDP and SDGs. NDP was an outcome of the Diagnostic Report identifying the contemporary challenges of the country while SDGs are set of targets to be achieved by addressing the challenges mentioned in NDP. Some aspects of SDGs (Table. 3.4) were not directly covered under the framework of NDP but the soul of SDGs is in harmony with the aims envisioned in the NDP.

NPC (2012) noted that 98 targets of SDG were directly covered by one or the other objectives of NDP while 44 were not related to NDP in any way. Out of 44 unattended targets of SDG by NDP, 32 targets were covered by other Government programme and 12 targets were out of the country's context. It shows that fully covered and partially covered targets of SDG by NDP account for 74% convergence reflecting the strong convergence between the two. As shown in Fig. 3.1, 57% SDG targets were fully addressed, 17% partially and 26% targets were not addressed.

The NDP prioritize capabilities and functioning to transform the economy and realize the desired change. It works on the interdependence between capabilities opportunity and employment and their combined impact on social changes and better living standard.

The detailed contribution of NDP objectives in the realization of SDGs are given in Table 3.5. It indicates the targets of SDGs covered by the objectives of NDP. Each NDP objective exhibit frequency varying from moderate to high level conducive to SDGs. Targets with higher multiplier effect need to be prioritized to attain the goals by 2030. The NDP objectives having direct influence and the potential to influence the SDGs are in first silo, objectives having moderate influence are in second while the third silo indicate little or non-influencer objectives.

Table 3.4 Aspects of SDGs

Sr no	
1	No one will leave behind
2	Sustainable development is the soul of all development programme
3	Social, economic and environmental issue are clubbed for sustainability in all aspects of humanity
4	Translate the economy in to growth and inclusive development
5	Ensure access of institution for all
6	New partnership for development

Source United Nations (2015)

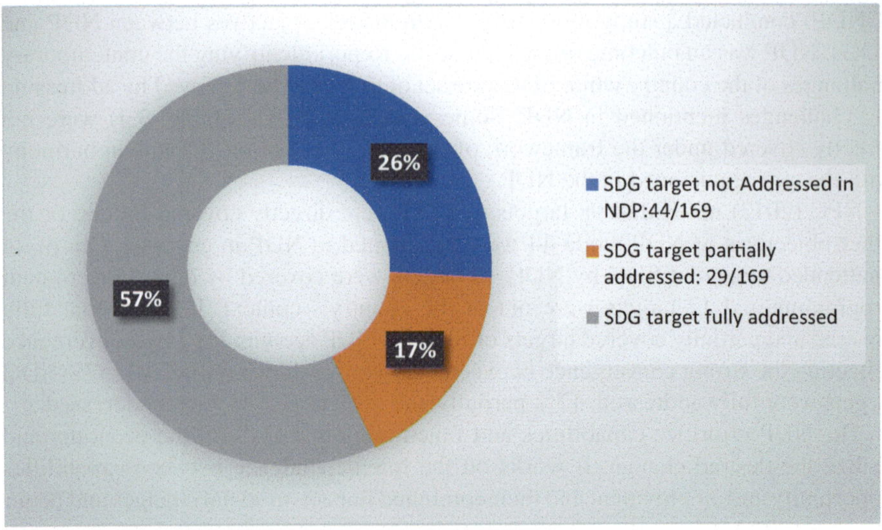

Fig. 3.1 Convergence of NDP in to SDGs. *Source* Department of Statistics, Republic of South Africa (2019)

Assessment of Sustainable Development Goals in South Africa

This chapter has identified goals from each dimension of SDGs, from social goals SDG1 (no poverty), SDG2 (zero hunger), SDG3 (health and wellbeing), SDG4 (quality education), SDG5 (gender equality), SDG6 (water and sanitation); from environment, SDG7 (Clean and affordable energy), and includes SDG8 (decent employment), and SDG10 (inequality) from economic goals for further discussion.

SDG1: No Poverty

The concept of poverty can be defined in many ways. Perhaps, it belongs to one of those concepts that generally elude definition. A person may be in poverty if they are unable to meet the necessities to support healthy life. (Mubangizi & Mubangizi, 2005). The alleviation of poverty depends on understanding poverty and its root cause and also the situation in which it operates. In South Africa, poverty is closely attached with race and gender (Soudien, 2022). Living conditions of blacks are worse than whites. Blacks comprises in huge proportion at the bottom of income and wealth pyramids, black women are most disadvantaged among them (OXFAM, 2020).

Table 3.5 NDP's level of contribution to SDGs

High SDG contribution	Medium SDG contribution	Low SDG contribution
The unemployment should reduce to 6% by 2030 from 24.9 in 2012	The proportion of adults in works force should reach 61% from 41%	The labor force participation should reach 65% from 54%
	The proportion in work force (rural) should increase to 40% from 29%	GDP increase by 2.7 time in real term, needed average annual GDP growth of 5.4 over the period
Raised the proportion of national income earned by bottom 40% should rise to 10% from 6%	Intraregional trade in Sothern Africa Should increase to 25% by 2030 from 7% now	GDP/capita increase to R110000/person by 2030 from R50000/person in 2010
Ensure access to clean, potable water for drinking and availability of water to agriculture and industries	Trade with neighbor increase to 30% from 15%	Expansion of ownership to historically marginalized cohort
	Availability of competitively priced broadband	Access to the power grid should rise by at least 90% by 2030
	Waste disposed to landfill should be reduce	By 2030, country would require an additional 29,000 MW electricity. Almost 10900 MW of current capacity is to exhaust, anticipate new build of more than 40,000 MW
	Renewable energy of 20,000 MW should be contracted	Share of renewable should be at least 20,000 MW
Well prepared indicator for natural resources with health publication to identify resource to inform policy	Balancing the trade between primary and processed agriculture	Reduce the water demand in urban areas to 15% by 2030
Better preparedness against extreme climate	Strong and spatial planning system well integrated across the wing of Government	Expand the public transport system. Make them environment friendly, cheaper and integrated
Investment in modernization of agriculture should increase, expansion of commercial expansion to protect the rural livelihood	South Africa should increase trade with neighboring countries t0 30% from 15%	Increase the amount of land and ocean under protection
	Strong and spatial planning system well integrated across all the departments of Government	
South Africa should create 643,000 direct and 326,000 indirect job opportunity in the agriculture and allied sector	Better facility of public transport	Attain the peak, plateau and decline trajectory for greenhouse gas emissions with peak being reached around 2025

(continued)

Table 3.5 (continued)

High SDG contribution	Medium SDG contribution	Low SDG contribution
Formal settlement for all informal settlement on well located land by 2030	Reverse the spread of communicable disease	An economy wise carbon price should be entrenched
Secure society for citizens of South Africa by 2030	Deployment of health workers to provide cares to families and communities	Zero emission building by 2030
	Decent living standard regardless of their income	Bridging the distance between home to office
	Fixation of minimum social floor to live	More job in or outskirt of urban township
Police should well equipped, well trained, professional to guard the marginalized and vulnerable without any discrimination	Children should benefit from the social programme (nutrition, health, education, social care and safety)	Take the longevity to 70 years
State intervention for development	Provide social security like income support to unemployed and other market related incentives	Reduce IMR and MMR
Public service should be protected from political interferences	Equal opportunity to all citizen	Fill the post with competent candidate
Staff should be provided with all amenities (authority, competence and support) to perform their jobs		Reduce the cases of accident and violence to 50%
Better coordination between national-provincial-local Government to manage inter-Governmental system		Adress the issue of stunting, malnutrition and micronutrient deficiency among children
		Meet the challenges of skill deficit in social welfare sector
Transparent Government structure and political stability and enable public enterprises to achieve their development potential		Make life secure after retirement. State should provide guideline to make pensions safe and sustainable
Accountable and corruption free Government		Ensure social protection system, promotion of self-employment and mechanism to cover the risk associated with such
		Provide an effective social welfare system that produce better result for vulnerable groups

Source National Planning Commission (2021)

South Africa's National Poverty Line

SDG1 calls for reducing the poverty level in men, women and children as defined by country's national level. The national poverty line in South Africa categorize poverty on basis of household consumption including both food and non-food components. South Africa has created three silos of poverty namely food poverty line, upper bound poverty level, and lower bound poverty level. South Africa recognized the fluctuation in cost of goods and services as well as consumption pattern of households over time as a key element in estimation of national poverty line.

It is presumed that both factors change over period of time with different intensity. Goods and services costs change more quickly and frequently than the change in consumption pattern of a household (Fukase & Martin, 2020). Any change in cost of living is directly related to purchasing power of an individual or value set by the poverty lines (Carr, 2021). Thus, construction of poverty lines relies on two elements, namely (i) inflation's-adjusted mean index and (ii) new poverty lines. Table 3.6 illustrate South Africa's national poverty lines (inflation adjusted) for the period 2006 to 2020.

The new poverty lines are dynamic and change over the period of ten years or if any notable change occurs in consumption pattern of households. The recent change in national poverty lines was done in 2011 based on the IES 2010/11. According to 2014/15 Statistics of South Africa, 55.5% people lives in upper poverty (ZAR 992); 40% people lives in lower bound poverty (ZAR 662); 25.2% people lives below food poverty line (R441) (STATS SA, 2018).

According to Statistics of South Africa (2023), the progress in poverty reduction is almost stagnant during 2015 and 2020, upper-bound poverty level increased by 0.5 during the same period while lower-bound remained constant at 40%. However, trends in food poverty line shows very tepid progress of mere 0.7% (see Fig. 3.2).

Table 3.6 Trends of inflation adjusted poverty lines in South Africa (in ZAR)

Years	Food poverty line	Lower-bound poverty line	Upper-bound poverty line
2006 (March)	219	370	575
2009 (March	318	456	709
2011 (March)	335	501	779
2015 (March)	441	647	992
2020 (March)	585	840	1268
2023 (April)	760	1058	1558

Source Department of Statistics, Republic of South Africa (2023)

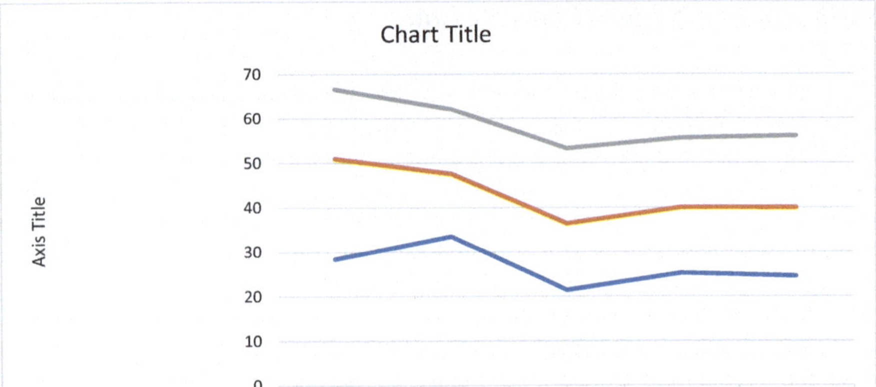

	2006	2009	2011	2015	2020
Food Poverty Line	28.4	33.5	21.4	25.2	24.5
Lower Bound Poverty Line	51	47.6	36.4	40	40
Upper-bound Poverty line	66.6	62.1	53.2	55.5	56

Fig. 3.2 Trends in inflation adjusted poverty lines (%), 2006–2015. *Source* Department of Statistics, Republic of South Africa (2023)

International Poverty Lines

Like national poverty line, international poverty line ($1.90 per day) shows similar trends. A significant progress in poverty reduction was registered during 2006 and 2011 but the success has been reversed during 2011 to 2015 and remained almost stagnant during 2015–2020 (see Fig. 3.3).

In September 2022, the World Bank has updated the international poverty line on basis of change in purchasing power parity. Therefore, the new benchmark for people in extreme poverty at $2.15 (2017 price) a day replaced the older threshold of $1.90 poverty line (World Bank, 2022a, 2022b, 2022c, 2022d). As per recent World Bank data, nearly 20.5% people in South Africa are living below the threshold of $2.15 per day consumption.

Multidimensional Poverty in South Africa

South Africa has made impressive progress in reducing the proportion of people living in multidimensional poverty. Similar to the above discussed trends of monetary poverty, the country has successfully lowered the share of multidimensional poverty between 2001 and 2011 from 17.9% to 6.3%. While the pace of reduction was slowed after 2011, it has not reversed till 2021 (UN, 2023a, 2023b, 2023c). Figure 3.4, demonstrate the trends of multidimensional poverty in South Africa.

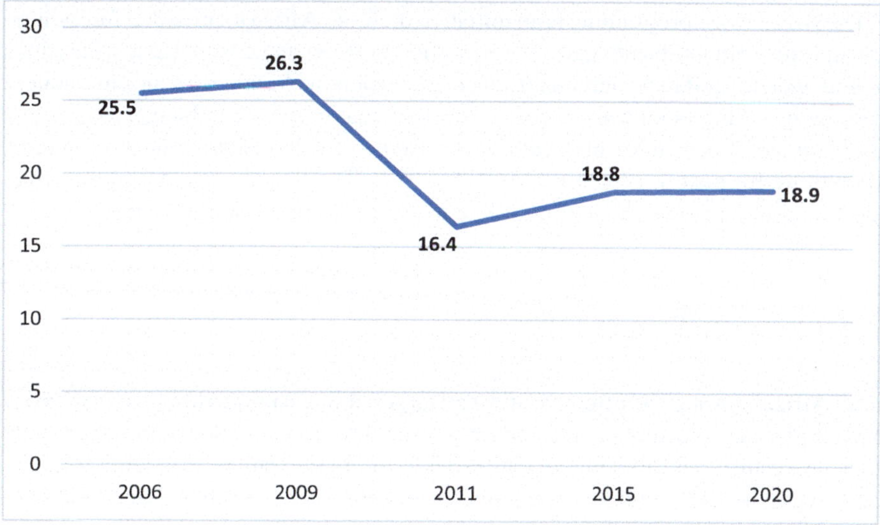

Fig. 3.3 Trends in International Poverty line (US$ 1.9/day). *Source* Department of Statistics, Republic of South Africa (2023)

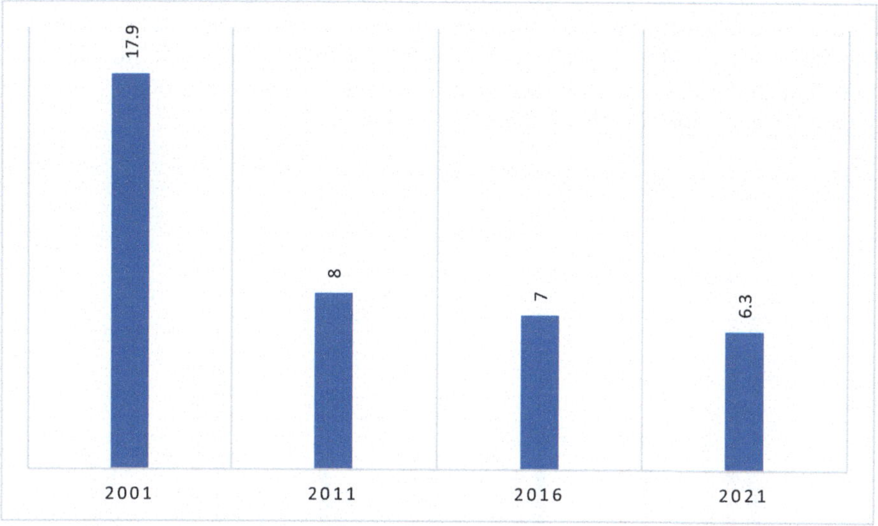

Fig. 3.4 Trends of multidimensional poverty 2001–2021. *Source* UN (2023a, 2023b, 2023c); Department of Statistics, Republic of South Africa (2023)

The social wage programme is an initiative of South Africa to provided free education and free primary health care, access to free housing and ensure basic necessities such as water, sanitation and electricity). According to South African government news agency, 'the social wage' scheme received massive share in budget to promote disadvantaged households and vulnerable groups, 59.2% of the total non-interest expenditure for next three years (SAGNA, 2022). The Social wage programme had a profound impact on reducing multidimensional poverty in South Africa.

Social Security System

South Africa's strong and effective and well targeted social protection system (SPS) is an exemplar in protecting people and promoting development. Social grants provide financial security and to some extent provide the main source of income to many South Africans. IMF readings noted that grants are well targeted and unconditional to employment, its ratio is 1: 3 (one individual out of three people) (Miyajima, 2023). Despite being an upper middle-income country, South Africa's share of SSS is quite high among the peers. South Africa spends 15.4% of total government expenditure on SPS, accounting for 3.3% of GDP. Nearly half (51%) of the SPS goes to the old age pension, and 38% dedicated to child support grant (World Bank, 2021). Expansion of social security system played a crucial role in improving people's life. ILO data shows that all the children's families are covered under social protection scheme and are equal beneficiaries in other indicators (see Fig. 3.5).

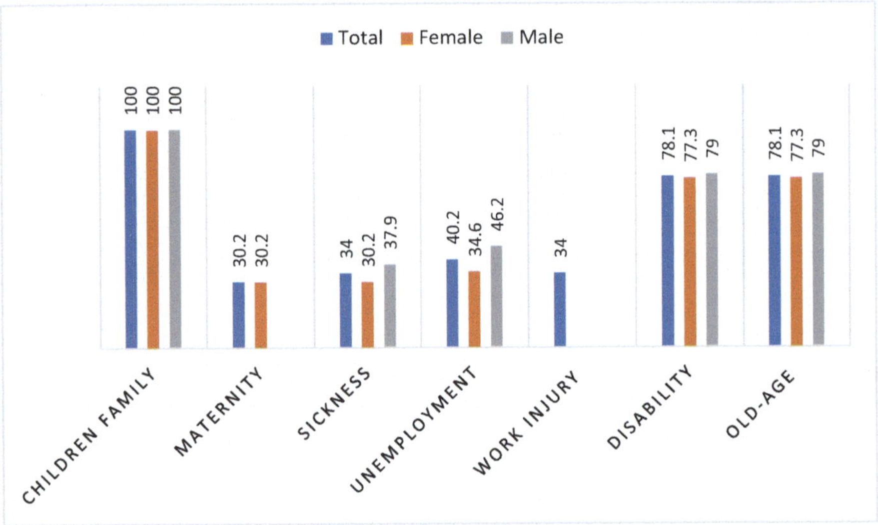

Fig. 3.5 Effective Coverage of social protection. *Source* International Labor Organizations (2023)

Leila Patel's study finds that nearly two out of three households are beneficiary of social grant in South Africa and almost 60% of the country's workforce is protected by social insurance (Patel, 2021). Grants have lowered the poverty rate ranging between 10–40% depending on definition of poverty and significantly bridged the inequality gap.

SDG2: Zero Hunger

SDG 2 is committed to end hunger, make a food secure world, improve nutrition intake, and promote sustainable agriculture pattern. Undernourishment is considered as one of the major challenges in achieving the full potential of an individual. It also plays a major role in pushing people to poverty. Undernourishment coupled with poverty are deadly and have the potential to ruin the present as well as the future of developing countries, especially in the continent of Africa (Adeyeye et al., 2017). Studies have established that poverty is strongly linked with childhood malnutrition (Rahman et al., 2021; Reyes et al., 2004). A malnourished child in a poverty ridden family have a grim chance of escaping poverty and is more likely to entrap in the vicious cycle of poverty with potentially chronic health consequences poverty (Jord & Monks, 2021).

Poverty and undernourishment are intrinsically linked. Thus, the South Africa share of undernourished people has shown an increase with the incidence of poverty (FAO, 2024). As shown in Fig. 3.6, the proportion of undernourished people (three-year average) almost doubled during 2006/08 to 2020/22. At the same time the counts of undernourished people rose from 1.8 million in 2006/08 to 4.7 million in 2020/22, with a mere decrease between 2013/15 and 2016/18.

Child Poverty and Malnutrition in South Africa

World Health Organization (WHO, 2020) revealed that almost half of the children in South Africa live in poverty. South Africa has a huge proportion of hunger and undernourished children. An estimated 30% of children live in family with poor access to healthy food, and unemployed adults. Every third child (under age of five) is stunted which amounts to 1.5 million. Every second child death is linked to chronic malnutrition among children. Nearly 75,000 children are suffering from obesity.

As illustrated in Fig. 3.7, trends in malnutrition namely stunting, overweight, and wasting is almost stagnant with some fraction of fluctuations. During 2015 to 2017, very small change was observed in stunting as it increased by 0.4, from 22.4 in 2017 to 22.8 in 2022. Over the same period, the proportion of overweight children was reduced by 0.8, from 12.9 to 12.1 and wasting remained stagnant during the same period (FAO, 2024).

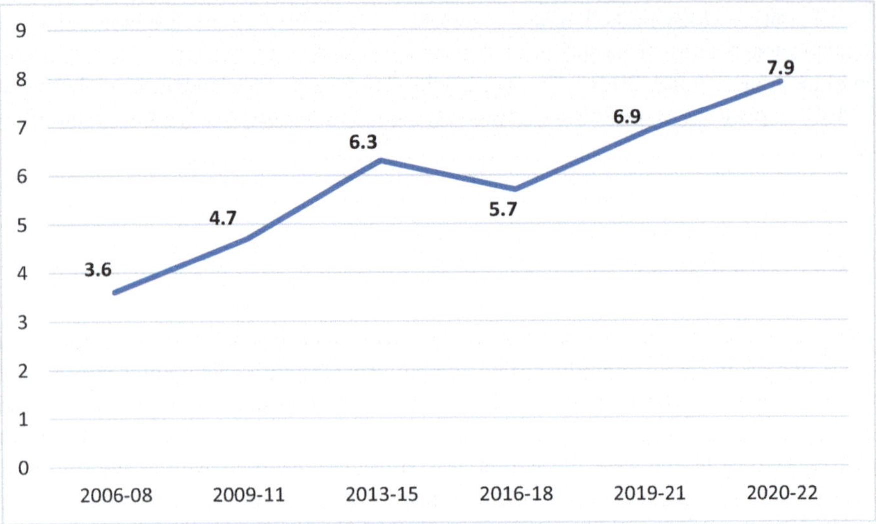

Fig. 3.6 Prevalence of undernourishment 2006–2022. *Source* FAO (2024)

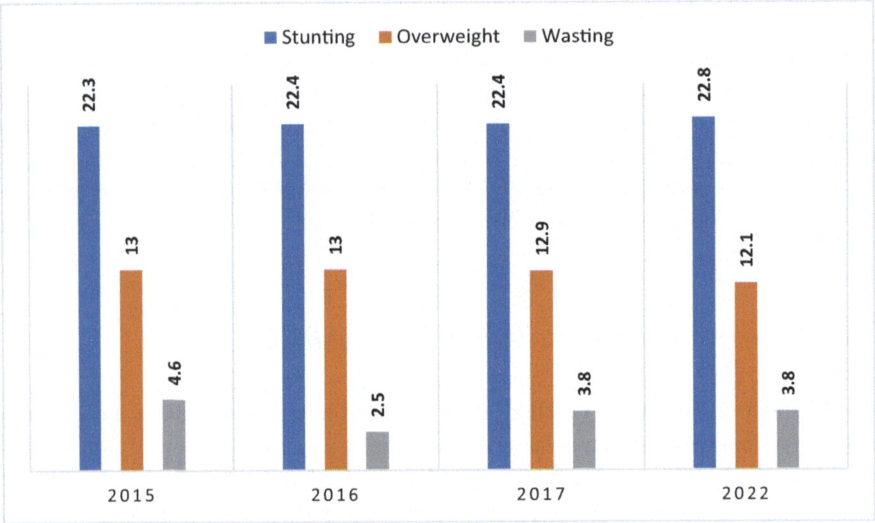

Fig. 3.7 Trends in malnutrition (2015–2022). *Source* FAO (2024)

The trend of malnutrition was uneven across the provinces of South Africa. UNICEF (2020a, 2020b) literature revealed that in 2016, three provinces namely Gauteng (34.2), Free State (33.5), KwaZulu Natal (28.5) surpassed the nation average (28.5) in stunting, whereas Eastern Cape (24.8), Limpopo (21.9), Mpumalanga (21.9), Northern Cape (21.4) and North West (22.9) fall below the national average,

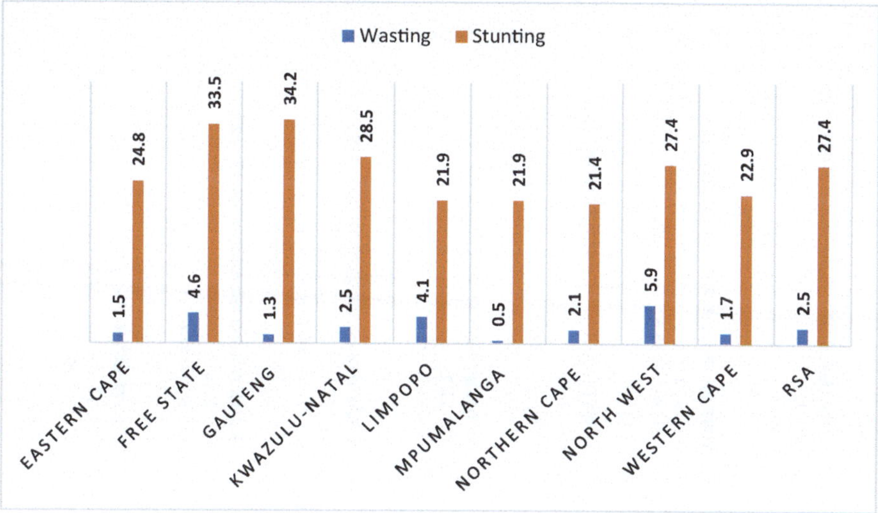

Fig. 3.8 Trends in Malnutrition at Regional Level, 2016. *Source* UNICEF (2020a, 2020b)

Western Cape (27.4) share the same rate with national average. Data of wasting shows similar trends (See Fig. 3.8).

SDG3: Health and Wellbeing

Section 27 of the Constitution of South Africa ensures health care and services including reproductive health and the National Health Act leverage the commitment made in the constitution by providing the framework of health infrastructure in South Africa. These mandates help South Africa towards attaining SDG3.

SDG3 is dedicated to promote overall health and wellbeing of all. SDG3.1 and SDG3.2 are committed to bring down maternal mortality rate (MMR) to 70 per 100,000 live births, neo natal death rate to 12 per 1000 live births, and under five mortality rate (U5MR) to 25 per live births (UN, 2016). Figure 3.9, indicate progress in MMR, NMR and U5MR. During 2015 to 2020, MMR reduced to 127 per 100,000 live births from 141 per 100,000 live births. Over the same period U5MR reduced to 33.7 deaths per 1000 live births from 37.3 deaths per 1000 live births. Country has achieved the targets of NMR in 2011 with 11.7 deaths per 1000 live births (UNICEF, 2023). Ongoing trend indicates that South Africa likely to achieve the target of MMR and U5MR.

The solution of poor health care lies in Universal Health Care (UHC) services. SDG 3.8 prioritize UHC for all. According to news agency Reuters, nearly 15% of South Africa are able to afford the comprehensive private healthcare system, while huge proportion of the poor, especially blacks, queue for hours at public hospitals

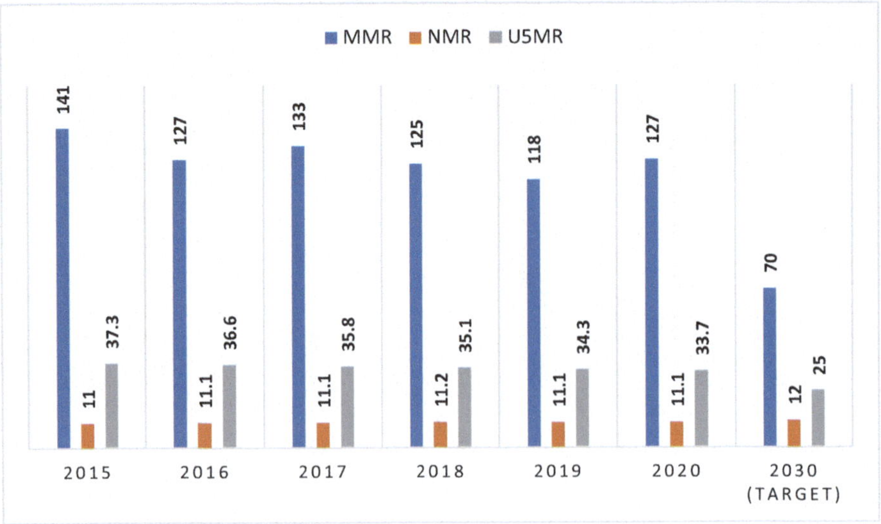

Fig. 3.9 Trends in MMR, NMR and U5MR (2015–2020) *Source* UNICEF (2023)

(Reuters, 2023). In a key note address, Joe Phaala, Minister of Health, Republic of South Africa, admitted the inadequate health infrastructure in country (OIS SA, 2023). In order to improve the access to healthcare services, government has launched national insurance in line with UHC mandate given by the United Nations.

In 2016, Demographic and Health Survey of South Africa reported that birth attended by skilled health worker increased to 96.7% from 84% in 1998 (NDoH, 2016). Institutional deliveries have a positive impact on maternal mortality and infant mortality. South Africa has achieved the target of neonatal mortality rate. The cases of HIV (15–24 year of age) were reduced to 1/1000 in the period of 2012–17 from 2.8/1000 uninfected people in the period of 2002–2005. The corresponding figure for age group 15–49 is 1/1000 and 2.8/1000 respectively, as reflected in Fig. 3.10.

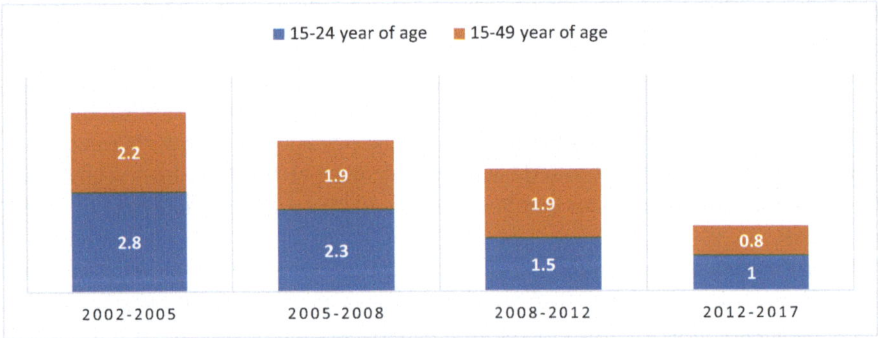

Fig. 3.10 HIV prevalence in South Africa. *Source* SABSSM 2012, SABSSM 2017, HSRC

Alcohol consumption which negatively affects other health targets has been a serious concern for Government due to its increasing trends. According to estimates, consumption of liquor increased to 7.27/capita in 2017 from 7.13/capita in 2011 (SAWIS 2011–17).

SDG4: Quality Education for All

Education is identified as a major challenge for sustainable development in the Diagnostic Report (2012). NDP has prioritized the challenges in its objective for Vision 2030 to achieve **SDG4**. The Government has developed a unique sector-specific framework known as Action Plan for 2019 and 2024 (anticipated) to strengthen the basic education sector.

South Africa has shown impressive progress in keeping pupils in school along with the successful completion of their grades. The Department of Basic Education, South Africa (2023) shows that at the primary level the completion rate reached 96.6% in 2021 from 73% in 2002. The country has registered continuous increase in the share of learner who completed grade7, grade 9, and grade 12 as depicted in Fig. 3.11.

South Africa has achieved high rate of secondary school pass-out with 94.2% in 2021 from 73% in 2002. The corresponding figure for senior secondary was 61.3% during the same period.

The improvement in pre-university education also translated in the progress in access to tertiary education. The Department of Higher Education and Training, Republic of South Africa in its report recorded the increasing trend in enrollment in post school education during 2002 and 2021, from 9.2% to 14.6%. Consequently,

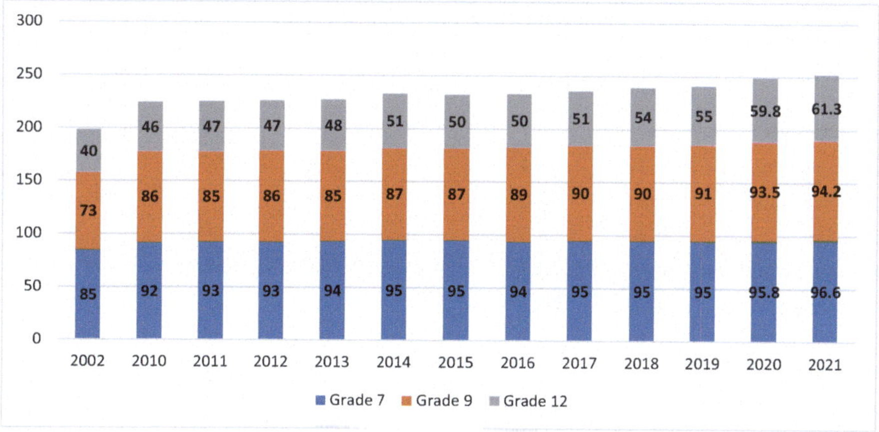

Fig. 3.11 Proportion of children who completed Grade7, Grade 9 and Grade 12 (%). *Source* Department of Basic Education, Republic of South Africa (2023)

the share of people without any school education descended to 3.2% from 11.4% over the same period.

Progress at the national level resonate in improvement at the provincial level, though unevenly. Table 3.7 shows the students completion rate in Grade 12, Grade 9, Grade 7 at the provincial level. Western Cape registered remarkable progress in completion grade 12, the share increased to 73.5 in 2021 from 51.3 in 2015, while Limpopo stood at bottom with 51%. At grade 9, again Western Cape retained the top position with 98.7% completion rate and Northern Cape and North West position in bottom rung with 86.3%. At Grade 7, Gauteng with 98.2 completion rate stood at the top position.

The Organization for Economic Co-operation and Development (OECD) observed that in South Africa the chances of employability increase with the progression in degree (OECD, 2019). However, despite significant progress in enrollment in tertiary education, South Africa still lags behind in gross enrollment ratio (GER) in comparison to other upper-middle income countries, including China, Colombia, Malaysia and the lower-middle income countries such as India, Philippines, and Algeria, during 2017 to 2020 (Stats SA, 2023a, 2023b, 2023c, 2023d, 2023e, 2023f). As shown in Fig. 3.12, South Africa with GER in tertiary 27.2% stood far behind China (58.4%), Columbia (52.2%), Algeria (52.5%) and Malaysia (42.6%). In contrast, South Africa's GER in tertiary education was quite closer to India (29.4%) and above Ghana (18.7%).

In line with the traditional education system, progress in enrollment in non-tertiary education (skills development and reskilling) for both employed an unemployed people was observed in South Africa. The country has adopted public–private collaboration model in Post School Education and Training (PSET) programme in order to promote skill education in South Africa. In 2020, a total 343 PSET institutions (85 public and 258 private) were exclusively running skill education courses. There

Table 3.7 Trends of pupil's completion (%) at provincial level (2015 and 2021)

	Grade 12		Grade 9		Grade 7	
	2015	2021	2015	2021	2015	2021
Western Cape	51.3	73.5	85.9	98.7	96.3	97.9
Eastern Cape	32.1	46.4	77.4	89.8	89.9	97.1
Northern Cape	43.1	50.8	73.4	86.3	87.8	91.1
Free State	48.8	58.5	85.6	94.3	95.6	91.3
KwaZulu-Natal	50.7	60.6	90	94	94.8	96.3
North West	40.7	64.1	80.5	86.3	91.8	93.9
Gauteng	63.4	67.1	92.7	97.8	98.3	98.2
Mpumalanga	49.4	62.7	85	94.2	93.8	96
Limpopo	37.7	51.4	87.8	91.8	95	97

Source Department of Statistics, Republic of South Africa (2023)

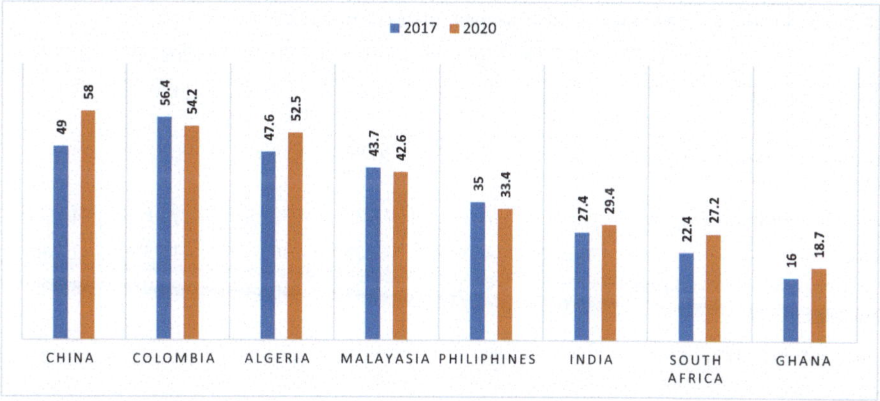

Fig. 3.12 Trends in GER in tertiary education (2017–2020). *Source* Department of Statistic, Republic of South Africa (2023)

are nearly 2 million students in PSET institutions, from which 84.3% students are enrolled in public institutions while 15.7% are enrolled in private institutions.

SDG5: Gender Parity

Gender disparity has a deep impact on the lives of children living in Sub-Saharan Africa (Baten et al., 2021). UNICEF literature observed a close association between the gender of a child and the possibilities of realizing their full potential, with traditional biased gender norms (UNICEF, 2020a, 2020b).

In South Africa, women constitute 51.1% of the population against 48.9% men. The movement for gender equality in South Africa traces its root in the protest against the Apartheid Pass Law. On 9th August, 1956, 20,000 women marched to the Union Building of Pretoria to deliver a letter of dissent against the oppressive apartheid regime's Pass Law, designed to control and reduce women to submission at the mercy of their counterpart (Panchia, 2021).

The post-apartheid government recognized the rights of women and ensured gender parity **(SDG5)** in article 3 of chapter 9 of the Constitution of South Africa: '*The state may not discriminate of base of race, gender sex, pregnancy, marital status, sexual orientation, age, disability, religion, conscience, belief and language and birth*'. Thess Constitutional provisions provides a base for policy legislation aimed at promoting gender equality in South Africa.

The government's commitment for women's transformation is reflected in several laws dedicated to their empowerment encompassing different sectors of society. For example, Promotion of Equality and Prevention of Unfair Discrimination (PEPUD) Act to ensure equality and eliminate biasness against women, Employment Equity

Table 3.8 Incidence of violence against women (2016/17 to 2022/23)

	2016/17	2017/18	2018/19	2019/20	2020/21	2021/22	2022/23
Contact Sexual Offenses	1488	1221	1254	1179	1059	1130	859
Attempted Sexual offences	2073	2066	2146	2076	1800	2027	2376
Sexual Assault	6271	6786	7437	7749	7025	7798	7483
Rape	39,828	40,035	41,583	42,289	36,330	41,739	42,780
Total Sexual Offences	49,660	50,108	52,420	53,293	46,214	52,694	53,498

Source SAPS (2023)

Act promotes equality at workplace, and victims of domestic abuse finds support from the Domestic Violence Act (OIS, 2023).

However, despite the various measures for safeguarding women at every level of the society, sexual offences against women have increased after registering a decline between 2019/20 (53,293) and 2020/21 (46,214). The increase in sexual offences has also been attributed to COVID-19 lock down. As shown in Table 3.8, except contact sexual offence all other indicators show an increasing trend.

Unpaid Work in South Africa

Recognition of unpaid care work has been advocated by SDG5.4. Unpaid work is a key element of economic activity and welfare of an individual. It includes caring of child and elder peoples, and household chores. A significant part of the day is consumed in such forms of unpaid work which is not recognized by any method of GDP accounting. On an average, woman across the world works three hours more than men, the gap differs across the countries (Seedat & Rondon, 2021).

The persistent gender inequality in unpaid work call for data on time use. Redressal of unpaid work needs focused policy design including gender equity. In 2000, three countries of South Southern Africa namely, South Africa, Madagascar, and Mauritius launched a time use survey. However, the ten-year cycle was only repeated by South Africa in 2010, while Madagascar and Mauritius stopped the survey in 2001 and 2003 respectively (Wekwete, 2017).

World Bank data reveled that during 2010–2019, women in South Africa spend 2.4 times as much in comparison to men in unpaid domestic work. In 2010, women's 15.6% time was consumed in unpaid work while men spent 6.5% of their time in unpaid work over the same period (World Bank, 2024a, 2024b). An IMF meta study on gender and unpaid work held that by improving female labor force participation a country can bridge the gender gap in total hours of unpaid work (Alonso et al., 2019).

The gap in unpaid work in South Africa, Mexico, and United Kingdom reduced by 20%, while in Pakistan it reduced by 45%.

Equal Participation of Women at All Levels of Decision Making

South Africa made an astonishing progress in expanding the share of women in National Parliament, from 33% in 2004 to 46% in 2019. The country has achieved parity in cabinet ministers, from 42 to 50% over the same period (Maluleke, 2023a, 2023b).

However, the representation of women in decision making at the national level was missing at the local level government. As depicted in Fig. 3.11, males occupied more mayoral position than females. In 2021 men held 68% mayoral seats while women occupancies stood at 32%. The share of occupancy differs in the provinces with only Northern cape achieving gender parity. Free State (47.8%) is close to achieving equality while other provinces fell below 40%.

SDG6: Clean Water and Sanitation

UN General Assembly resolution (64/292) recognized that water and sanitation is vital in realization of all human rights. Way back in 1996, the Constitution of South Africa provided the right of water in Sect. 27(B). The Constitution paved the way for policy formation for water. Water Service Act (NWA) 1997 provided regulation for water and sanitation services in the country. SDG6 call for clean drinking water and sanitation services for all.

SDG 6.1.1 is committed to ensure clean and safe drinking water service to all households. General households report (2023) contends that the progress in supply of clean and safe drinking water to households has decreased during 2021–2022, from 83.6% to 83.4%. But if we compare the 2022 figure with 2015, it shows an increase of 3.4 points.

However, another government report (2023) revealed the presence of pollutants in water contamination in water supply system (WSS) of South Africa. The total number of WSS in South Africa was 958, out of which 442 (46%) received unacceptable pathogen water quality status, while remaining 467 (49%) and 49 (5%) received excellent and good water quality status (see Fig. 3.13).

Further, the report indicated that the consumption of polluted water is not safe for drinking purpose since it has chronic health implications. As shown in Fig. 3.3.14, 423 (44%) and 229 (24%) WSS received unacceptable status for drinking purpose due to presence of heavy chemical.

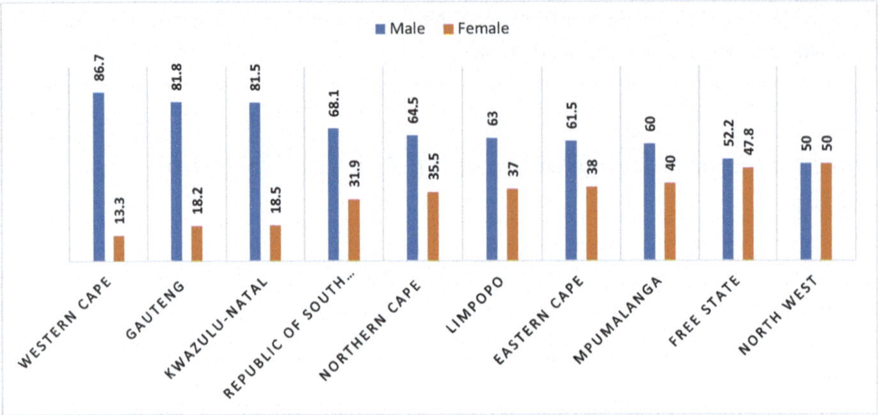

Fig. 3.13 Mayoral seats held by Sex in Provinces of South Africa in 2021. *Source* (Stats SA, 2022)

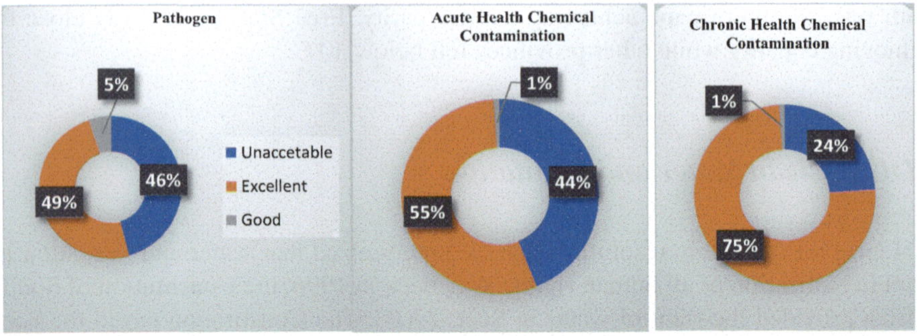

Fig. 3.14 Proportion of contamination in Water Supply System. *Source* Department of Warter and Sanitation (2023)

Table 3.9 depicts the distribution of water Supply System among the provinces of South Africa. Registering the presence of 59% pathogen in water supply, Eastern Cape stood at top position in bad quality of water while western cape with 22% is positioned on top in supplying good quality water.

Access to Sanitation Services

WHO report (2023) on progress and sanitation in Africa finds that 31% people have access to safe and managed sanitation services while 28% used basic sanitation facilities in 2022. The report further indicated the glaring gap in access to sanitation services between rural and urban of Africa. In urban Africa, 2 out of 3 person lack safe and managed sanitation services while the corresponding in rural area was 3 out

Table 3.9 Distribution of Water Supply System among the provinces

	WSS #	Pathogen * (%)	Unacceptable #	Unacceptable * (%)
Eastern Cape	154	86.49	91	59
Free State	80	76.52	45	56
Gauteng	29	98.7	6	21
KwaZulu-Natal	172	92.62	70	41
Limpopo	84	81.6	18	21
Mpumalanga	100	71.08	64	64
Northern Cape	176	72.26	104	59
North West	39	88.13	17	44
Western Cape	124	96.89	27	22
Total	**958**	**84.93**	**442**	**46**

Source Department of Warter and Sanitation (2023)

of 4. During the same period half of the urban population lacks basic hygiene where as in rural Africa the proportion was 4 out of 5.

In line with SDGs, South Africa's National Sanitation Policy-2016 is committed to promote access to sanitation services in the country. While South Africa has registered the progress in access to safe and managed sanitation service and basic sanitation facilities between 2015 and 2022, from 66.5% to 71.7% and 5.7% to 5.9% respectively, a significant gap still exists (UNICEF, 2024). Motsoeneng (2022) noted that metropolitan areas of South Africa performed better than national average in access to sanitation services.

In South Africa nearly 65% of the population used waterborne sanitation like flush toilet attached to a sewer tank or septic tank and may be connected to conservancy tank. 19% used improved ventilated pit latrines while 13% used pit toilets without ventilation pipes. Other remaining group either depend on compost or chemical toilets. About 1% still practiced open defecation in 2021 (Arumugam-Nanoolal, 2022).

SDG7: Clean and Affordable Energy

South Africa is going through serious problem of power shortage with some parts of the country experiencing 12-h power cut a day. Load shading in South Africa started in 2007 and have multiplied intensely reaching to a daily average of 9 h in 2022 (World Bank, 2023a, 2023b, 2023c, 2023d). Arwen Kazak (2023) in her study finds root of energy shortage in corruption, apartheid legacy, low investment, or mismanagement. South Africa has the potential to harness renewable energy which is also reflected in its policy design.

South Africa is a highly industrialized economy, so it has a big energy sector. As per the Energy Transition Index (ETI) ranking 2023, South Africa stood at 82nd position out of 120 countries. The ETI score improved between 2014 to 2023 by 6% (WEF, 2023). Consequently, the country has also registered improvement in system performance score, reinforced by improvement in access to energy and clean and affordable cooking fuels. Country's reduction in energy intensity has also led to advancement in the progress of sustainability in energy sector. Despite the historical challenges in energy transition, country has improved the share of renewable energy in the last ten years.

Sunshine is a notable feature of South Africa's climate (Fitchett & Hoogendoorn, 2018) but the country still lags behind in production of solar energy in comparison to countries with low sunshine. South Africa's share of solar power in the country's total energy mix is just 2.5%, while Germany's solar power contributes ten percent in total energy mix of country (Winkler, 2021). 80% of the country's energy is derived from coal, while the share of clean energy is almost thirteen percent, resulting in high intensity of carbon in total energy mix (Ziady, 2023).

South Africa is the only country in the African continent with a nuclear power plant. The country's two nuclear reactors produced 10,089 GWh energy in 2022, which amounts to 4.9% in the total energy mix (IAEA, 2023). However, both nuclear plants met the end of project life in 2024 (IRP, 2019). In order to enhance the capacity and production of nuclear energy, the country is working to add new mix of large nuclear power plants and small modular reactors combined in power generation capacity of 2500 MW in its nuclear profile.

As shown in Fig. 3.15, the total annual production and capacity of renewable energy in South Africa almost tripled during 2015–2023, from 2241 GWh to 6233 GWH over the same period. This includes onshore wind, concentrated Solar (3 h storage capacity), photovoltaic power, small hydro, biomass, and landfill gas. Onshore wind energy is the largest contributor in total renewable energy source, scoring nearly triple growth during 2015–2023, while concentrated Solar (3 h. storage capacity) remained stagnant at 500 GWh during 2019 t0 2023 (REDIS, 2024).

SDG8: Decent Employment

SDG 8, call for decent employment for all working hands. Studies shows that South Africa's unemployment is strongly linked to skill mismatch in job market (Habiyaremye et al., 2022a, 2022b). The official unemployment data shows that unemployment in South Africa rose from 22% in 1994 to 25% in 2014 and reached 31.9% in the third quarter of 2023. Going by the expanded definition of unemployment, it was 41.2% in the same quarter.

As shown in Table 3.10, unemployment in five provinces namely Eastern Cape (38.8%), Free State (38.5), Gauteng (33.7), Mpumalanga (35.6), North West (38.6) were above the national average while Western Cape (20.20), Norther Cape (20.2) and Limpopo (30.8) managed to keep their unemployment below national average

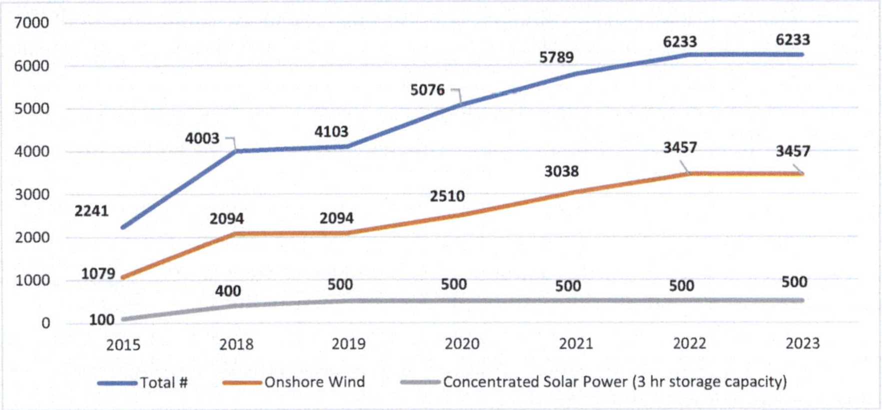

Fig. 3.15 Annual production and capacity of renewable energy in South Africa, 2015–2023 (in GWh). #Total renewable energy includes Onshore wind + Concentrated Solar + Photovoltaic Power + Small Hydro + Biomass + Landfill gas. *Source* Renewable Energy Data and Information Service (2024)

in Quarter 3 of 2023. Over the same period Labor Force Participation Rate (LFPR) shows similar trends.

Rising unemployment is a key challenge in post-apartheid South Africa. Throughout the 1970s, the country experienced low rate of unemployment. Unemployment rate slightly picked up in the last two decades of apartheid regime. The biased feudal capitalism lowered the opportunities for blacks and coloured while cushy socialism promoted whites (The Economist, 2019). The unemployment rate at the time of political transformation was lower than it is today (STATS SA, 2016a,

Table 3.10 Provincial unemployment and LFPR in South Africa, Quarter 3 2023 (%)

Provinces	UR	EUR	LFPR
Eastern Cape	38.8	43.9	52.9
Free State	38.5	44.7	61.1
Gauteng	33.7	39.4	68
KwaZulu-Natal	29.4	44.7	52.4
Limpopo	30.8	45.1	55.6
Mpumalanga	35.6	46.7	59.4
Northern Cape	26.3	42	53.4
North West	38.6	51.2	55.3
Western Cape	20.2	25.6	68.6
RSA	31.9	41.2	62.2

Source Department of Statistics, RSA (2023)
UR: Unemployment Rate, EUR: Expanded Unemployment Rate, LFPR: Labor Force participation Rate, RSA: Republic of South Africa

2016b). The historically disadvantaged communities are more vulnerable in the job market, especially among females, uneducated youths, and people living in rural areas.

IMF literature noted that the restructured economy of the post-apartheid era has intensified the economic growth along with positive growth in employment, but growth not kept match with the growing labour force. The rate of growth in labour force was nearly four time stronger than the rise in employment (Arora & Ricci, 2006). Banerjee et al. (2007) concluded that the rising unemployment in post aprtheid regime is due to the increasing demand of skilled labour particulary in mining and agriculture sectors and also the influx of black women in the labour market. Kerr and Martin (2020) in their work on post-apartheid labor market agreed with the rising unemployment in post-apartheid era but they also indicate the stagnant rate of employment over the same period.

The growing demand for skilled labour and large supply of unskilled labour happens simultenously. Consequently, more suply and less demand of labour force gives forward push to unemployment in South Africa. This is expected as formal employment enjoys lion share in total emplpyment of South Africa. In 2023, nearly 69.4% of people were employed in the formal sector, whereas only 18.3% work in infromal secotrs and the remaining 6.7% work in private household and 5.7 in agriculture sector (STATS SA, 2023a, 2023b, 2023c, 2023d, 2023e, 2023f).

The key challenge of unemployment in South Africa lies within its education system. President, Cyril Ramaphosa in a conference address stressed that the country's education need to revamp in order to produce skilled workforce needed for sustainable future (Business Tech, 2021).

The most salient feature of South Africa's labor force is the chronic mismatch in the demand and supply of labor force. The country is griped under surplus supply of unskilled labor against the surge in demand of skilled labor. Curtailing this mismatch is a huge task for policy makers. The country can succeed in accomplishing the target of decent job for all through focused skill education.

As shown in Fig. 3.16, group of workers below the secondary education are more vulnearble in the job market. In the second quarter of 2023 the unemployment rate among the people below secondary level was 38.3 which incresed to 38.8% in third quarter of the same year, higher than the national unemployment rate. However, unemployment was least among the people with garduate dergee followed by other tertiary certificate/diploma.

The majority of unemployed people are characterized under the specific group 'not in education, employment and Training (NEET)'. Nearly one-third of youth (15–24 years) fall under this specific category. Like other disadvantaged groups, persons identified in NEET belong to rural and disadvantaged areas with limited employment opportunities. The large size of NEET in south Africa calls form urgent policy intervention.

Post School Education and training (PSET) in South Africa is key to promote post school education and training apart from traditional approach of basic education to train the increasing number of NEET. It also provides opportunities of reskill and upskill to those workers who lost their livelihood during COVID-19 pandemic.

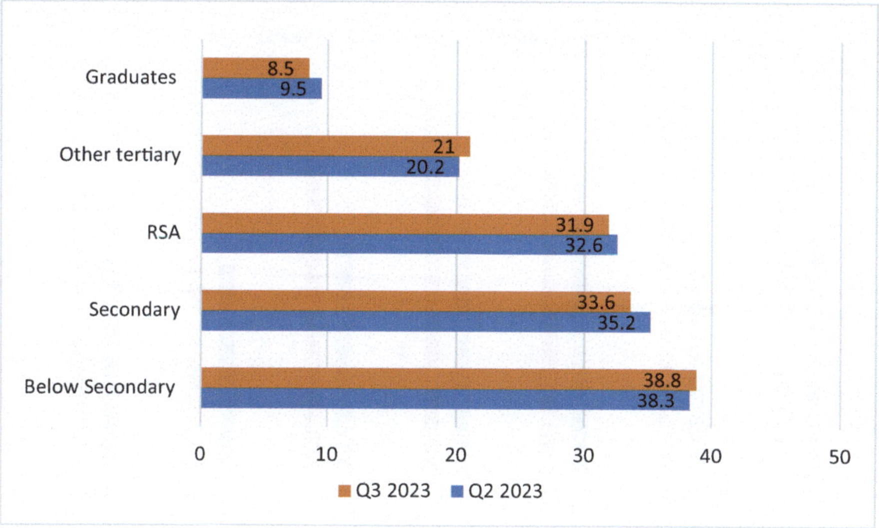

Fig. 3.16 Level of qualification and Unemployment in South Africa, Q3 2023 and Q2 2023. *Source* Department of Statistics, RSA (2023)

Further, PSET also accommodate those students who failed to complete their education along with those who never attended any schools. However, the gap between people with NEET and enrollment in PSET is continuously increasing. As shown in Fig. 3.17, the number of people in NEET increased from 3,044,000 to 3,432,000 in 2021 while during the same period enrolment in PSET reduced from 2,473,000 to 2,249,000.

The experience of middle-income countries shows that unemployment rise with decline in subsistence agriculture. Consequently, the informal sector emerges as a new hub for low skilled employment, in a poor working condition. Furthermore, the demand for low skilled job tends to decline due to advancement in automation which needs more skilled labor (Patel et al., 2018). The situation is more prevalent in Sub-Sharan Africa, particularly in South Africa where majority of youth are not employable due to low skill sets.

UNESCO report (2023) suggest that a better Technical and Vocational Educational Training (TVET) system will enable countries to invest more in skills and training of youths in order to reap the benefit of demographic dividend. In South Africa, TVET institutions are considered as key element to tackle the rising unemployment. TVET is a key constituent of Post School Training in South Africa. TVET colleges are providing skill education and training to the South African youth, improving their competencies in job market, and enabling them to secure decent jobs. Most of the TVET colleges are managed by government, with a target to expand the TVET system by 2030. However, a recent world Bank Report (World Bank, 2024a, 2024b) shows that insufficient fund is the major hurdle in further expansion and improvement

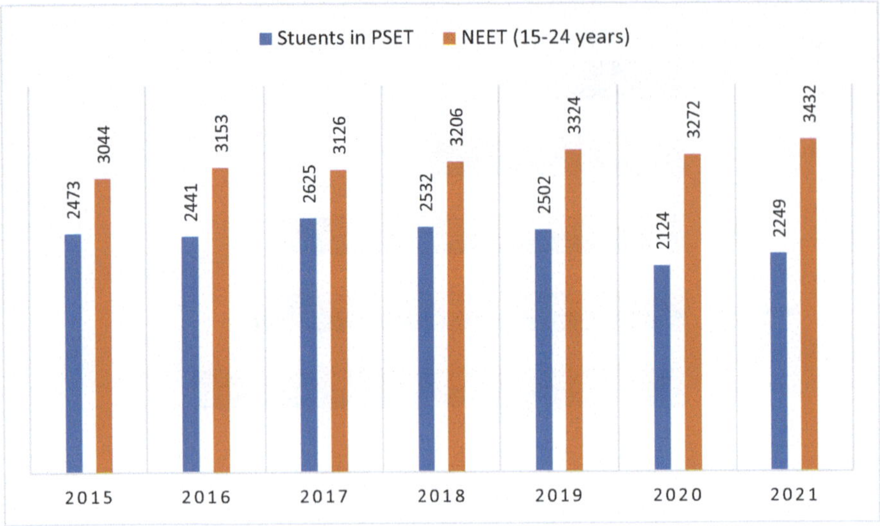

Fig. 3.17 Number of Students in PSET vs NEET (15–24 years) (in 1000). *Source* Department of Higher Education and Training, RSA (2023)

of the quality of TVET in South Africa. An estimated 80% of allocated funds are utilized in covering employee expenses.

Since 1994, the government of South Africa has implemented various policy initiatives to create more jobs to reduce the rate of unemployment. The spillover impact is also anticipated in reducing poverty and inequality in country. However, the poor quality of education at the primary level obstructs the way of future learning. The combined effects of poor schooling and limited access to tertiary education in South Africa lowers the chances of employment, especially among the youth. ILO (2020) data reveled that in 2018, merely 1% of South Africa's youth are enrolled in vocational training.

Entrepreneurship: A Way to Sustainable Future

Entrepreneurship is a key to decent jobs and economic growth. Launching a business is never an easy task, but once started it can yield significant benefits to the people at the individual level as well as national level. Micro Small and Medium Enterprises (MSME) are crucial in reducing unemployment, covering an estimated share of two-third of the total jobs in an economy (Jobanputra, 2023).

OECD (2022) data shows that South Africa possess nearly 2.6 million MSME, around 37% qualified in formal, another 54% are micro-enterprise, and remaining 15% are established in rural areas. Another international agency reported MSME in South Africa share 98% of formal business accounting nearly 2 million companies and registered 2% growth in 2022 (UN, 2023a, 2023b). It shares in total employment of South Africa was nearly 60% and its contribution to GDP was approximately 34% (IFC, 2019). The share of MSME in total value addition of South Africa increased from 18% in 2010 to 40% in 2020. Still, South Africa has lower base of entrepreneurship in comparison to other African nations. South Africa should have three times stronger base of early-stage entrepreneurship with its current trends of per-capita GDP.

MSME ownership also exposes the existence of socioeconomic inequality in South African society. Most of the micro industries are owned by blacks and males. Men have better chances of starting a new venture than women. Global Entrepreneurship Monitor (2023) indicate that unemployed youth took MSME as a solution of unemployment. The scarcity of jobs motivates them start their own business, and also make difference in the world. By doing this, they are anticipating reliable source of income to build wealth. However, COVID-19 disturbed MSME growth and the expectation of new jobs is declining. The creation of additional employment with new businesses or from running businesses are declining. Three-out of-ten businesses anticipate to add another six jobs in 2019 which reduced to just two-out of-ten.

South Africa has recognized the global technological advancement and its link to MSME. The concern to promote robust, innovative and agile entrepreneurs has been envisaged in the country's masterplan for the next ten years in National Integrated Small Enterprise Development (NISED). International Finance Corporation (IFC) channeled $3 billion funds to strengthen the entrepreneurship in South Africa, the plan is committed to create 50,000 new jobs in next five years. Additional 73,000 and 98,000 jobs would be created in collaboration with private sector lenders.

Public Work Programme

In response to the rising unemployment, South Africa has conceptualized Expanded Public Works Program (EPWP) in 2003, a job creation programme initiative. The EPWP was launched to reduce poverty by encouraging employment opportunities (ILO, 2016). EPWP mandate covers all unemployed, unskilled, poor and vulnerable population of South Africa. The jobs provided in EPWP are for short period, generally in between six to twelve months. One person can claim 230 days of work annually.

The programme was built on earlier public work programme such as Zivuseni, Zibambele and working for water. These jobs are mostly characterized in labor intensive jobs. EPWP targeted to create nearly 90,000 jobs in its first year, followed by 170,000 jobs every year by third of its launch (DPWI, 2003).

According to EPWP phase four business plan report (2019), the first phase (2004–2009) of EPWP was launched with a target of 1 million job opportunities by the target year. To uplift disadvantaged group, 55% jobs were reserved for women, 40% for youth and two percent for physically disabled people. In Second phase (2009–2014) nearly four million job opportunities were created against the target of 4.5 million job opportunities. Third Phase (2019–2024) was launched in line with international labor law. The objective is to provide social safety net of job opportunity and income support, delivered though public asset and services. Till 2019, 4.5 million job opportunities were created against the target of six million job opportunities. In financial year 2021/22, EPWP has created 1,016,646 jobs opportunities out of which 40.1% (415,897) was occupied by youths (18–35 years) (OIS, 2022).

EPWP has made a significant contribution in poverty alleviation. It promotes inclusive development in two ways, allowing participant's access to education and benefiting them with stipend, in return, participants establish food gardens which enhance communities to access food. (Mohapi, 2016). EPWP empower participants, provide them opportunity to live with dignity and is also a way to promote social inclusion for marginalized communities (Scully & Moyo, 2021). However, the success of EPWP is not sustainable because it is a temporary relief and have no such effect on removing total poverty (Maphanga & Mazenda, 2019). Glen & Mearn (2021) argued that the desired outcome of EPWP is yet to realized.

SDG10: Reduce Inequality

United Nations literature noted that inequality has intensely widened across the world, becoming a serious threat to our common future (Guterres, 2020). SDG 10 is committed to reduce inequality. South Africa belongs to the world's oldest custom union, Southern African Custom Union (SACU), and also the biggest country among the member countries namely Botswana, Namibia, Eswatini, and Lesotho. World Bank Report identified SACU as the most unequal region and South Africa as a highly unequal country among the world Bank ranking of 164 countries based on global poverty database 2022 (Sulla, Zikhali, & Facundo, 2022).

The persistence of inequality is one of the salient features of South Africa. To some extent, inequalities are associated with economic growth and may be for short duration. It can result from the uneven reciprocation of people towards the ongoing growth. It becomes a severe threat to the country when majority of people experience economic and social retardation.

Inequality in South Africa is also linked to racial segregation, a curse of apartheid regime and colonialism. South Africa's Inequality Report (2019) held that policy interventions in post-apartheid regimes have shown tepid progress, especially income inequality which has widened after 1994 (Valodia, 2023). World Bank records indicates high value consumption expenditure Gini Coefficient of 0.67 for South Africa in 2018 (World Bank, 2023a, 2023b, 2023c, 2023d).

Living Condition Survey 2014/15 shows the glaring inequality among the provinces of South Africa. Soley Gauteng share ZAR 0.4 in every ZAR spent in South Africa. Over the same period male headed household received income of ZAR 138,168, while the corresponding figure for female headed household was ZAR 98,911. Further report shows stark spending gap between White-headed household and Black-headed household. The Black's household spent one-fifth of white's household who spent triple of national average. (STATS SA, 2022).

World Inequality Report (2022) shows that Income gap varies across the African continent, from 13 to 15% in Mali, Ethiopia, Guina Nigeria and from 40 to 63% in Namibia, Central African Republic, Zambia and South Africa. The income before taxes of bottom 50% of the population in South Africa is 63 folds lower than the income received by top 10% population, whereas after taxes the gap reduced to 24 folds (Chancel et al., 2022). IMF (2020) estimation shows that bottom 40% earns 7% of the country's income, 16% lower than other emerging countries.

Like income inequality, South Africa also tops the wealth inequality. As shown in 3.17, the top ten percent people in South Africa hold 85% of wealth while in United States the figure varies in between 70–75% while in UK the share remains only to 55%. South Africa is also experiencing high wealth inequality with low chance of intergenerational mobility. Evidences indicates severe disparity in wealth inequality among South Africans. Study at Witwatersrand revealed that 0.01% population (3500 people) holds nearly 15%, the top 0.1% holds 25% of country's wealth. In terms of net wealth, average wealth of top 1% is ZAR 27.8 million (nearly $ 944,000), while the corresponding figure for bottom 50% -ZAR 16,000 negative wealth possession indicates liabilities more than assets (Chatterjee et al., 2023). Further, the study contends that there is no change in wealth inequality since 1994 (Fig. 3.18).

South Africa aims to eradicate poverty and contain inequality by 2030. The government has realized the potential of country's demography to accelerate the pace of inclusive development. Government of South Africa has used redistributive policies to contain the persistent inequality including fiscal redistribution policy. Fiscal policy has potential to reduce inequality. The government focused programme such as expansion of social spending, targeted financial support, promotion of entrepreneurship among the victims of apartheid, diversifying wealth possession, are measures to ensure economic justice.

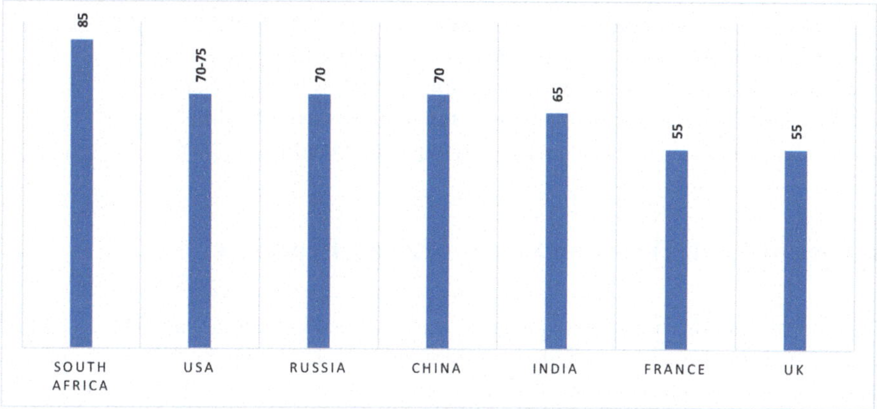

Fig. 3.18 Wealth inequalities among the various countries. *Source* Chatterjee, Czajka, and Gethin (2023)

Emergence of Covid-19 Pandemic in South Africa

The first positive case of novel coronavirus in South Africa was registered on 5th March 2020. The infection was detected in KwaZulu Natal, carried by a South African citizen who travelled to Italy (OIS, 2020; Nordling, 2020). Moonsar, et al., (2021) observed that South Africa became the most heavily infected country in African Continent within a short duration between March and August.

Giandhari and et al. (2021) noted that the spread of the virus has been highly uneven across the provinces. The first incidence of mortality was reported from KwaZulu Natal and Gauteng. Initially, most of the mortality were reported from KwaZulu Natal, due to the large influx of patients in private hospitals located in Durban, and later replaced by Western Cape. In the urban cohort of Gauteng, the infection was limited among the rich and elites who travelled across for holidays, especially among the people living in Sandton area. Consequently, low mortality was reported over time. On the other side Western Cape experienced intense spread of coronavirus especially in Cape Town, a major tourist hub of South Africa.

In the absence of preventive vaccine to control the rapid expansion of the contagious virus, the world turned towards the nonpharmaceutical strategies, so did South Africa. In the first move, the operation of educational institutions, restaurant and bars were stopped. Followed by complete nation-wide lockdown effective from midnight of 26th March for next 21 days. On 27th March the incidence of mortality passed the mark of 1000.

South Africa introduce an 'alert levels' model of lockdown framed in a scientific manner, stringent confinement in highly infected area with ease in containment measure with decline in infection and improvement in health system as well (DEH, 2020). It also includes the degree of preparedness of country's health system. The five alert levels were recognized ranging from alert: -

- Alert level 5 (rapid spread of virus with poor health system)
- Alert level 4 (High level of infection and moderate health system)
- Alert level 3 (moderate level of infection as well as health system)
- Alert level 2 (moderate level of infection and improved health system)
- Alert level 1 (low infection and highly equipped health system).

Impact of Lock Down on South African Economy

Notwithstanding the impressive progress in social and economic sector, the yield of development instruments has failed to address the triple challenge of Poverty, unemployment, and inequality since the political transformation of 1994. The COVID-19 pandemic made the situation worse as containment measures snatched the gains achieved by continuous process in the well guided framework. South Africa imposed strict containment measure to save their population from the COVID 19 pandemic.

Containment measure cost three million jobs and the income source dried for additional 1.5 million workers (Smit, S. 2020, CRAMS, 2020, STATSSA, 2020).

The pandemic unmasked the biggest challenge of South Africa, its unemployed labor force. The high unemployment and the large size of inactive work force were badly affected by the crisis. A telephonic survey in sixth week of the lockdown revealed that every fourth person had experienced income drop (Seekings & Nattrass, 2020). The impact of crisis on low wage workers or daily wagers was four times higher than high-income employee (Fengler et al., 2021).

COVID 19 exposed the failure of the existing infrastructure in South Africa. The crisis had various adverse impacts on potential utilization of various sectors in South Africa which led to significant contraction of gross fixed capital generation in the first quarter of 2020, in caparison with the same period of previous year. The country witnessed recession in pre COVID 19 years as seen in Fig. 3.19. During crisis, policy makers suggested the augmentation of Government spending to accelerate the economy to realise full employment and curb unemployment because of contraction in aggregate demand. Altman (2020); Arndt et al., (2020); Naidoo (2020) estimated the decline in GDP to 7.2% for 2020, flattening the fiscal deficit to 15.7% of GDP. Although the country was in recession before the outbreak of pandemic in 2020 (STATSSA, 2020).

Lockdown's estimates indicate sudden and massive impact on the economy of South Africa. The shutdown of industries contracted the income with differential distribution. Study of Arndlt, et al. (2020) contends that wage earning dropped by 30% whereas gross operating surplus reduced by 40%. Subsequently, GDP (factor cost) declined by 34%.

The pandemic has expanded the share of people without adequate food in last seven days, from 20% in 2020 to 25% in early period of the crisis (Wet-Billings, 2023). The stringent containment measures put the country on standstill as all economic activities were stopped except essential services for almost four months.

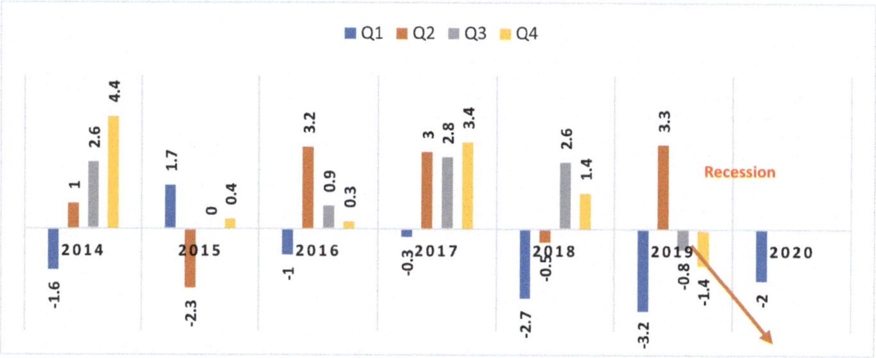

Fig. 3.19 GDP Trends in South Africa, 2014–2020 (all quarters). *Source* Department of Statistics, Republic of South Africa (2021)

National Income Dynamic Survey (2020) revealed that business closure due to lockdown resulted in job loss, an estimated 10% male ad 22% females' workers becoming unemployed.

The motivation for investment in infrastructure to meet the challenges of recession is rooted in economic models as well as empirical literatures investigating the link between fiscal policy and output response. Studies suggest that expansionary fiscal policy have the potential to stimulate the aggregate demand (Sinai, 2014); (Spilimbergo et al., 2009); (Verick & Islam, 2010); (Auerbach & Gorodnichenko, 2012): (Singh, 2023). Infrastructure investment is assumed as growth amplifier due to its durability and multiplying effects. IMF literature indicate the strong positive link between fiscal expansion and economic recession over the economic expansions (Baum et al., 2012). In the context of current fiscal expansions, fiscal stimulus can be more fruitful if the funds are invested in infrastructure rather than tax cut or public consumption (Batini et al., 2014).

In the wake of the pandemic, the Government of South Africa announced 'extraordinary coronavirus' budget to wage war against COVID 19. To sharpen the 'instrument of development', South Africa announced R500 stimulus package equivalent to 10% of the country's GDP. Stimulus involved enormous money, injected to build sustainable infrastructure and prepare country to reap the benefit of 4th industrialization. Health was the first priority and Government dedicated R20 billion for health infrastructure and service (see Table 3.11).

The stimulus is an immediate relief to accelerate growth and generate jobs in the transition phase of the crisis period and make economy more resilient to externalities. Sguazzin (2020) argued that 1.8, million jobs are expected from the said package after the normalization of health crisis.

Stimulus of ZAR 500 billion is an extra support from existing budget to leverage job opportunities with ZAR 200 billion for credit guarantee program and provision of ZAR 100 billion for job creation in small and medium sized industries and other informal business. The Government also dedicated ZAR 20 billion for immediate

Table 3.11 Sector wise distribution of Stimulus package, South Africa

	Amount (ZAR) Billion
Credit guarantee program	200
Support for informal business sector, SME for job creation	100
Provision for income support (tax deferral, extension of ETI and SDL holiday)	70
Grants for protecting vulnerable communities for six months	50
Wage protection	40
Health including other frontline services	20
Provision for support to municipalities	20
Total	500

Source Department of National Treasury, Republic of South Africa (2020)

relief to health sector. Provisions for wage protection and support for vulnerable communities were also included in fiscal response (see Table 3.11).

Sibeko and Isaacs (2020) study finds that in 2018, fiscal multiplier for South Africa was equal to 1.68 which is understandable, as fiscal stimulus of ZAR 1 billion would translate the economy in ZAR 1.68 billion with 6900 new job opportunities. Further, study rejected the idea that stimulus of ZAR500 billion (10% of GDP) would amplify economy over ZAR 800 billion with 3.5 million new job opportunities.

During the onset of COVID-19 pandemic, South Africa managed to mitigate the impact of the crisis on poverty level. The Government spent $6 billion on social safety net programme which benefitted almost 29 million people living in poverty. World Bank report (2022a) noted that the pandemic has created an increase in poverty levels across the world. At the regional level, huge surge was observed in Low Income Countries and Lower Middle-Income Countries, whereas Upper Middle Income Countries experienced contraction in extreme poverty, for instance, Brazil and South Africa. Further, report included that South Africa has set an example for reaching vulnerable households who suffered from income loss due to pandemic.

Challenges

The political shift in South Africa gave hope for equality, employment and poverty free country. The Government devised various instruments of development to get rid of poverty, skill mismatch, unemployment and inequality. Although the Government has lifted many from poverty, reduced inequality to some extent and make people food secure, the country is still the most unequal state in the world. Unemployment is almost 30% and poverty are all time high. High burden of disease is another challenge which the country continues to face. Accessibility to education is highly uneven. A key element of unemployment in South Africa is skill mismatch between demand and supply of labor force. Technically under-skilled laborers are not suitable for prevailing jobs.

Access to health and education is dependent on access to resource and to some extent possession of wealth. To uproot inequality, it is important to focus on wealth inequality along with other forms of inequality. Numerous studies have documented the extreme income inequality in South Africa. Primarily these studies are focused on wage and income inequality, through multiple dimensions in different time period by employing different data source (Alvaredo & Atkinson, 2010; Hundenborn & Jellema, 2019; Mosomi & Wittenberg, 2020; Gouzoulis et al., 2021).

The prime contributor of unemployment NEET is continuously increasing with minor reduction since 2022. As shown in Fig. 3.20, the share of NEET among the groups (15–60 years) and (15–24 years) has witnessed a decline since 2022.

Health is influenced by exogenous factors such as food security, education, employment, housing, water and sanitation, and energy collectively termed as Social Determinant of Health (SDH). Attainment of SDG health targets is directly proportional to improvements in SDH. Health sector need immediate attention as PHCs are

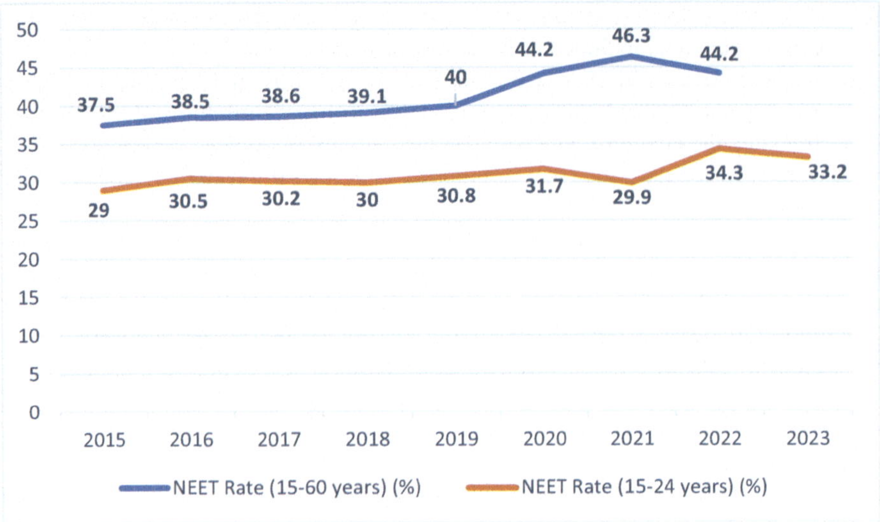

Fig. 3.20 Trends of NEET (15–60 years) and 15–24 years, 2015–2023. *Source* Department of Higher Education and Training, RSA (2023)

witnessing infrastructural challenges. Health sector workers are also not adequately skilled.

Legislation gaps pose a bigger challenge of Gender parity. South Africa drastically need a developed framework of international standard such as Women's Empowerment and Equality Bill to attain the targets of Gender parity.

Water challenges are a threat to others goals as it is considered as key enabler. South Africa is facing severe water crisis from population pressure, ecological pressure, and management of data and coordination challenges influencing the progress of goals.

Electricity shortage is another key challenge. It has negatively affected the daily life of people and also amplified their hardship. Persistent load shedding has reduced economic activity while costly diesel generator increases the production cost. It also disrupts the service delivery of health and education and other social infrastructure such as water and IT.

Mabunda et al. (2023) in their study pointed that the small businesses in South Africa are dysfunctional due to frequent load shading. Further, study shows that small businesses are losing their customers' trust and loyalty because they are not fulfilling their demands on time because of load shading. Subsequently, the desired socio-economic development targets such as poverty alleviation, inequality and unemployment are halted, and may be reversed if situation continue longer.

COVID-19 Pandemic and poor structural growth have intensified existing socio-economic challenges in South Africa. South African economy has returned to its pre-pandemic GDP but is still struggling to improve employment level. According to World Bank estimation, 74,000 people were not reemployed till the mid of 2023. It remains the world's most inequal country with more than half of its population living in poverty. The absence of entrepreneurship in school curriculum is a major challenge in entrepreneurship development in South Africa. Introduction of enterprise literacy at all educational level could bridge the skill mismatch and the job market.

From the time of Adam Smith critics have argued that the effectiveness of the invisible hand can only be realized in a structured framework. SDG country report (2019) identified lack of leadership, institutional weakness and low trust in Government as the major challenges faced by economic goal. Low foreign investments and sluggish growth rate constraint the attainment of SDGs.

Conclusion

The legacy of apartheid regime is yet to be fully addressed. Rising inequality, poverty and unemployment continues to be major hurdles in accomplishing the target of SDGs. The progress of SDGs targets is based on the experiences of MDGs. The National Planning Commission framed the National Development Programme-2030 and converged the objectives of NDP with the targets of SDGs. The South African Government successfully converged the 74% objectives of NDP with targets of SDGs. In the first 3-year South Africa was on the path of Attainment of SDGs but in late 2019 the country witnessed a recession and in 2020 the pandemic halted the progress and even reversed some of the hard-earned progress.

Although, South Africa has achieved significant progress in enrollment in pre tertiary schools, the quality of education is not up to standards. Because of this, the success at the primary level has failed to translate at the tertiary level. Furter, skill mismatch in labor markets draws immediate attention. TVET colleges are not up to international standard.

There is an urgent need to improve the quality of drinking, it could be serious threat to human health in the long term.

The data on income and wealth inequality is amply clear which also demands immediate attention. Economic development is closely linked with energy supply. Energy supply in South Africa is dominated by Eskom which has failed to maintain the balance between supply and demand of electricity. The situation has worsened since 2007 and people are forced to live with load shedding.

The pandemic has exposed the prolonged structural weakness that have been progressively deteriorating since the 2008–2009 economic crisis.

To shift South Africa on the path of sustainable development and inclusive economic growth in order to realize the vison of human development, the solution lies in human capital. The Covid-19 crisis has exposed the deficit and disproportional human capital across the country.

The country's strength lies in its large proportion of youth population. There is an urgent need for overarching investment to enhance the capabilities of youth and improve their chances of employability and help them in securing decent jobs. The Government has tried to revitalize the economy with stimulus package and the success of SDGs solely depends on the relief provided by the Government and policy formation in long term as discussed in the chapter.

References

Adeyeye, S. A., Adebayo-Oyetoro, A. O., & Tiamiyu, H. K. (2017). Poverty and malnutrition in Africa: A conceptual analysi. *Nutrition & Food Science, 47* No.6, Available at: https://doi.org/10.1108/NFS-02-2017-0027.

AfDB. (2022). *Africa industrialization index 2022.* Available at: https://www.afdb.org/en/documents/africa-industrialization-index-2022: The African Development Bank Group.

African Union. (2015). *Popular version.* Ethiopia. Available at: https://au.int/sites/default/files/documents/36204-doc-agenda2063_popular_version_en.pdf: African Union Commission.

Allen, C., Nejdawi, R., El-Baba, J., Hamati, K., Metternicht, G., & Wiedmann, T. (2017). Indicator-based assessments of progress towards the sustainable development goals (SDGs): A case study from the Arab region. *Sustainability Science and Implementing the Sustainable Development Goals, 12*, 975–989. Available at: https://doi.org/10.1007/s11625-017-0437-1.

Allen, C., Reid, M., Thwaites, J., Glover, R., & Kestin, T. (2020). Assessing national progress and priorities for the Sustainable Development Goals (SDGs): Experience from Australia. *Ustain Science, 15*, 521–538. Available at: https://doi.org/10.1007/s11625-019-00711-x.

Alonso, C., Brussevich, M., Dabla-Norris, E., Kinoshita, Y., & Kochhar, K. (2019). *Reducing and redistributing unpaid work: Stronger policies to support gender equality. IMF Working Paper No. WP/19/225.* Avaialbele at: mf.org/en/Publications/WP/Issues/2019/10/15/Reducing-and-Redistributing-Unpaid-Work-Stronger-Policies-to-Support-Gender-Equality-48688: International Monetary Funds.

Alvaredo, F., & Atkinson, A. B. (2010). Colonial rule, apartheid and natural resources: Top incomes in South Africa, 1903–2007. *Centre for Economic Research and Policy (CEPR), Discussion Paper No. 8155.*

Ansell, C., Sørensen, E., & Torfing, J. (2022). Translating global goals to local contexts. In C. Ansell, E. Sørensen, & J. Torfing, *Co-Creation for Sustainability* (pp. 41–56). Emerald Publishing Limited. Available on: https://doi.org/10.1108/978-1-80043-798-220220004.

Arndt, C., Davies, R., Gabriel, S., Harris, L., Makrelov, K., Modise, B., . . . Anderson, L. (2020). *Impact of Covid-19 on the South African economy. SA-TIED Working Paper 111.* Available at: https://sa-tied.wider.unu.edu/sites/default/files/pdf/SA-TIED-WP-111.pdf: UNI-Wider.

Arora, V., & Ricci, L. A. (2006). Unemployment and the Labor Market. In M. Nowak, & L. A. Ricci, *Post-Apartheid South Africa: the First Ten Years* (pp. 23–46). International Monetary Fund. Avaialable at: https://www.imf.org/external/pubs/nft/2006/soafrica/eng/pasoafr/sach3.pdf.

Arumugam-Nanoolal, P. (2022). Flushing toilets aren't the solution to South Africa's sanitation problem. *The Conversation*, pp. Available at: https://theconversation.com/flushing-toilets-arent-the-solution-to-south-africas-sanitation-problem-194768.

Auerbach, A. J., & Gorodnichenko, Y. (2012). Measuring the Output Responses to Fiscal Policy. *American Economic Journal: Economic Policy*, 1–27.

Banerjee, A., Galiani, S., McLaren, J. L., & Woolard, I. (2007). Why has unemployment risen in the new South Africa. *National Bureau of Economic Research Working Paper No. 13167*, Available at: https://www.nber.org/system/files/working_papers/w13167/w13167.pdf.

Baten, J., Haas, M. d., Kempter, E., & Selhausen, F. M. (2021). Educational gender inequality in Sub-Saharan Africa: A long-term perspective. *Data and Perspectives*, Available athttps://doi.org/10.1111/padr.12430

Batini, N., Eyraud, L., & Weber, A. (2014). A simple method to compute fiscal multipliers. *IMF Working Paper WP/14/93*, https://www.imf.org/external/pubs/ft/wp/2014/wp1493.pdf.

Baum, A., Ribeiro, M. P., & Weber, A. (2012). *Fiscal multipliers and the state of the economy. Working Paper No. 2012/286*. https://www.imf.org/en/Publications/WP/Issues/2016/12/31/Fiscal-Multipliers-and-the-State-of-the-Economy-40146: International Monetary Fund.

Bhorat, H., & Cassim, A. (2014). South Africa's welfare success story II: Poverty-reducing social grants. *Brookings*, pp. Available at: https://www.brookings.edu/articles/south-africas-welfare-success-story-ii-poverty-reducing-social-grants/.

Blanc, D. L., Liu, W., O'Connor, D., & Zubcevic, I. (2012). *Issue 1: Development cooperation in the light of sustainable development and the SDGs: Preliminary exploration of the issues. Rio+20 working papers*. Available at: https://sustainabledevelopment.un.org/content/documents/761workingpaper3.pdf: United Nations.

Business Tech. (2021). Young unemployed South Africans don't have the right skills: Ramaphosa. *Business Tech*, pp. Available a: https://businesstech.co.za/news/technology/526992/young-unemployed-south-africans-dont-have-the-right-skills-ramaphosa//.

Carr, S. C. (2021). Setting 'poverty thresholds': whose experience counts? *Sustainable Science*, 31–36. Avaiable at: https://www.ncbi.nlm.nih.gov/pmc/articles/PMC7451228/.

Chancel, L., Piketty, T., Saez, E., & Zucman, G. (2022). *World inequality report 2022*. Available at: wir2022.wid.world: World Inequality Lab.

Chatterjee, A., Czajka, L., & Gethin, A. (2023). *Working papers and research output*. Available at: https://www.wits.ac.za/scis/research-projects/wealth-inequality/working-papers-and-research-output/: Southern Centre for Inequality Studies., University of the Witwatersrand-hannes.

Cirillo, C., & Tebaldi, R. (2016). *Social protection in Africa: inventory of non-contributory programmes*. Available at: https://ipcig.org/pub/eng/Social_Protection_in_Africa.pdf: International Policy Centre for Inclusive Growth United Nations Development Programm.

Corder, C. K. (1997). The Reconstruction and Development Programme: Success or Failure? *Social Indicators Research, 41*(1/3), 183–203. Available at: https://www.jstor.org/stable/27522262.

DBE. (2023). *Annual Performance Plan: 2023/24*. Available at: https://www.education.gov.za/Resources/Reports.aspx: Department of Basic Education. Republic of South Africa.

DEH. (2020). *Disaster management act, 2002: Directions issued in terms of regulation 3(3) of the regulations made under section 27(2) of the disaster management act, 2002 (ACT NO. 57 OF 2002): Criteria to guide the determination of alert levels*. Available at: https://www.gov.za/sites/default/files/gcis_document/202008/43599rg11158gon867_1.pdf: Department of Health. Republic of South Africa.

Department of National Treasury. (2020). *Economic measure for Covid-19*. Department of National Treasury, Republic of South Africa.

DHET. (2023). *Persons who are not in employment, education or training (NEETS)*. Available at: https://www.dhet.gov.za/Planning%20Monitoring%20and%20Evaluation%20Coordination/Fact%20Sheet%20-%20NEETs_%20March%202023.pdf: Department of Higher Education and Training. Republic of South Africa.

DHS. (1997). *Urban development frame work*. Availabe at: https://www.gov.za/sites/default/files/gcis_document/201409/saurb.pdf: Department of Human Settlement. Republic of South Africa.

Doorgapersad, S. V. (2022). Overview of the implementation of Sustainable Development Goal. *Research in Business & Social Science. IJRBS, 11*(6), 2147–4478. Available at: https://ssbfnet.com/ojs/index.php/ijrbs/article/view/1889/1445.

DPWI. (2003). *Business plan for an expanded public works programme*. Pretoria. Available at: https://www.ilo.org/wcmsp5/groups/public/@ed_emp/@emp_policy/@invest/documents/projectdocumentation/wcms_asist_8163.pdf: DEPARTMENT OF PUBLIC WORKS and Infrastructure. Republic of South Africa.

DPWI. (2019). *Expanded public works programme (EPWP) Phase IV business plan 2019–2024*. Available at: http://www.epwp.gov.za/documents/Cross_Cutting/Monitoring%20and%20Evaluation/EPWP_PhaseIV_Business_Plan_2019 2024.pdf: Department of Public Works and Infrastructure. Republic of South Africa.

Drewes, J. E., & Aswegen, M. V. (2013). *National planning in South Africa: a critical review.* Available at: https://www.witpress.com/Secure/elibrary/papers/SDP13/SDP13016FU1.pdf: WIT Transactions on Ecology and The Environment, Vol 173.

DTI. (2007). *mplementation of Government's National Industrial Policy Framework: Industrial Policy Action Plan August 2007.* Available at: https://www.thedtic.gov.za/wp-content/uploads/IPAP_r1.pdf: Department of Trade and Industries.

DWS. (2023). *Blue Drop: Nationla Report 2023.* Water and Sanitation Department. Republic of South Africa.

Elizabeth Ninan Dulvy, T. Z.-M.-K., Kabubei, K. M., Matsebula, T., Monchuk, V., & Cunningham, W. (2023). *Investing in human capital in South Africa.* Available at: https://documents1.worldbank.org/curated/en/099111623064028345/pdf/P175444197fe220021bcb71d5549f85864b.pdf: The World Bank.

FAO. (2024). *FAOSTAT: South Africa.* Available at: https://www.fao.org/faostat/en/#country/202: Food and Agriculture Organization.

Fengler, W., Nelly, M.-F., Gill, I., Baduel, B., & Cuevas, F. (2021). South Africa after COVID-19 Light at the end of very long tunnel. *Brookings*, pp. Available at: https://www.brookings.edu/articles/south-africa-after-covid-19-light-at-the-end-of-a-very-long-tunnel/.

Fine, B. (2012). Assessing South Africa's New Growth Path: Framework for change? *Review of African Political Economy, 39*(134), 551–568. Available at: https://www.jstor.org/stable/42003209.

Fitchett, J., & Hoogendoorn, G. (2018). Does sunny South Africa really have an ideal climate for tourism? *The Conversation*, pp. Available at: https://theconversation.com/does-sunny-south-africa-really-have-an-ideal-climate-for-tourism-103852.

Fransis, D., & Webster, E. (2019). Poverty and inequality in South Africa: Critical reflections. *Development Southern Africa., 36*(6), 788–802.

Fukase, E., & Martin, W. (2020). Economic growth, convergence, and world food demand and supply. *World Development*,https://doi.org/10.1016/j.worlddev.2020.104954

GEM. (2023). *Global Entrepreneurship Monitor.* Available on: https://www.gemconsortium.org/reports/latest-global-report: Global Entrepreneurship Monitor.

Giandhari, J., Pillay, S., Wilkinson, E., Tegally, H., Sinayskiy, I., Schuld, M., … Fonseca, V. (2021). Early transmission of SARS-CoV-2 in South Africa: An epidemiological and phylogenetic report. *International Journal of Infectious Diseases*, 234–241. Available at: https://doi.org/10.1101/2020.05.29.20116376.

Glen, N., & Mearns, K. F. (2021). Expanded public works programme participants: Transitioning to tourism in South Africa. *Journal of Asian and African Studies*, 2014–2035. Available at: https://doi.org/10.1177/00219096211002875.

Gouzoulis, G., Constantine, C., & Ajefu, J. (2021). Economic and political determinants of the South African labour share, 1971–2019. *Economic and Industrial Democracy*, 184–207. Available at: https://doi.org/10.1177/0143831X211063230.

Guterres, A. (2020). Opening remarks at Nelson Mandela Lecture: "Tackling the Inequality Pandemic: A New Social Contract for a New Era". *Speaches*, pp. Avaialble at: https://www.un.org/sg/en/content/sg/speeches/2020-07-18/remarks-nelson-mandela-lecture-tackling-the-inequality-pandemic-new-social-contract-for-new-era.

Habiyaremye, A., Habanabakize, T., & Nwosu, C. (2022). Bridging the labour market skills gap to tackle youth unemployment in South Africa. *The Economic and Labour Relations Review, 33*(4), 786–805. Available at: https://doi.org/10.1177/10353046221138400.

Habiyaremye, A., Molewa, O., & Lekomanyane, P. (2022b). Estimating employment gains of the proposed infrastructure stimulus plan in Post-Covid-19 South Africa. *European Journal of Development Research, 34*, 540–567. https://doi.org/10.1057/s41287-021-00400-5

Haywood, L. K., Funke, N., Audouin, M., Musvoto, C., & Nahman, A. (2018). The sustainable development goals in South Africa: Investigating the need for multi-stakeholder partnerships. *Development Southern Africa, 36(5)*, 555–569. Available at: https://doi.org/10.1080/0376835X.2018.1461611.

Hundenborn, J., & Jellema, I. W. (2019). The effect of top incomes on inequality in South Africa. *International Tax and Public Finance, 26*, 1018–1047. Avaialbele at: https://doi.org/10.1007/s10797-018-9529-9.

IAEA. (2023). *Coutry profile: South Africa. Preamble and summary.* Available at: https://cnpp.iaea.org/public/countries/ZA/profile/preview: International Atomic Energy Agency.

IFC. (2019). *The unseen sector a report on the msme opportunity in South Africa.* Available at: https://www.ifc.org/content/dam/ifc/doc/mgrt/2019-01-msme-opportunity-south-africa.pdf: INternational Finance Corporation.

ILO. (2016). *South Africa—expanded public works programme.* Available at: https://www.social-protection.org/gimi/Media.action;jsessionid=uF1cuJLvyXAIQP87LzIJPt8slbHtT_jsiGO0zUQdWe1yMMvvQ-aM!1393577045?id=16462: International Labor Organizations.

ILO. (2020). *Global employment trends for youth 2020.* International Labour Organisation

ILO. (2023). *World social protection data dashboards.* Available at: https://www.social-protection.org/gimi/WSPDB.action?id=13: International Labour Organizations.

IMF. (2020). Six Charts Explain South Africa's Inequality. *IMF COUNTRY FOCUS*, pp. Available at: https://www.imf.org/en/News/Articles/2020/01/29/na012820six-charts-on-south-africas-persistent-and-multi-faceted-inequality.

IMF. (2023). *World economic outlook (OCTOBER 2023).* Available at: https://www.imf.org/external/datamapper/NGDPD@WEO/OEMDC/ADVEC/WEOWORLD: International Monetary Fund.

IRP. (2019). *Ntegrated Resource Plan (IRP2019).* Available at: https://www.energy.gov.za/IRP/2019/IRP-2019.pdf: Department of Energy. Republic of South Africa.

Jobanputra, K. (2023). Entrepreneurship: The Engine Of Growth Driving Our Economy. *Forbes*, pp. https://www.forbes.com/sites/forbesbusinesscouncil/2023/07/31/entrepreneurship-the-engine-of-growth-driving-our-economy/?sh=70c483a45af9.

Jord, J. O., & Monks, A. (2021). Food poverty and youth work—A community response. *Critical Social Policy, 42*(1), 64–84. Available at: https://doi.org/10.1177/0261018321996534.

Kerr, A., & Wittenberg, M. (2020). The post-apartheid labour market series. *Research Data Journal for the Humanities and Social Sciences*, 40–47. Available at: https://doi.org/10.25828/gtr1-8r20.

Kozak, A. (2023). Shedding the load: Power shortages Widen Divides in South Africa. *Insight: Kleinman Center for Energy Policy.* University of Pennsylvania, pp. Available at: https://kleinmanenergy.upenn.edu/news-insights/shedding-the-load-power-shortages-widen-divides-in-south-africa/.

Liu, Y., Huang, B., Guo, H., & Liu, J. (2023). A big data approach to assess progress towards Sustainable Development Goals for cities of varying sizes. *Communications Earth & Environment, 4*, Article number: 66, Available at: https://www.nature.com/articles/s43247-023-00730-8.

Mabunda, M. V., Mukonza, R. M., & Mudzanani, L. R. (2023). The effects of loadshedding on small and medium enterprises in the Collins Chabane local municipality. *Journal of Innovation and Entrepreneurship, 12*, Article number: 57, Available at: https://doi.org/10.1186/s13731-023-00327-7.

Maluleke, R. (2023a). *Gender Series Volume IX: Women Empowerment, 2017–2022.* Available at: https://www.statssa.gov.za/publications/Report-03-10-26/WE_Gender%20Presentation%2018-10_22_%20Final%20for%20SG%20.pdf: Statistics South Africa. Republic of South Africa.

Maluleke, R. (2023b). Messege: STATISTICIAN-GENERAL, MR RISENGA MALULEKE. In S. SA, *Sustainable Development Goals: Country Report 2023* (pp. 7–8). Department of Statistics. Republic of South Africa. Available at: https://www.statssa.gov.za/MDG/SDG_Country_report.pdf.

Maphanga, M., & Mazenda, A. (2019). The effectiveness of the expanded public works programme as a poverty alleviation strategy. *Administratio Publica, 27*(3) , 7–25. Available at: /https://repository.up.ac.za/bitstream/handle/2263/71497/Maphanga_Effectiveness_2019.pdf?sequence=1.

McGrath, S., & Akoojee, S. (2007). Education and skills for development in South Africa: Reflections on the accelerated and shared growth initiative for South Africa. *International Journal of Educational Development, 27*(4), 421–434. Available at: https://doi.org/10.1016/j.ijedudev.2006.07.009.

Miyajima, K. (2023). The link between social grants and employment in South Africa. *IMF: Selected Issues Papers, 2023*(39), Available at: https://doi.org/10.5089/9798400243493.018.

Mohapi, B. J. (2016). The social sector of the Expanded Public Works Programme as a strategy to alleviate poverty amongst vulnerable groups in Gauteng. *Development Southern Africa, 33*(5), 644–657. Available at: https://doi.org/10.1080/0376835X.2016.1204227.

Moonasar, D., Pillay, A., Leonard, E., Naidoo, R., Mngemane, S., Ramkrishna, W., … Modiseny, M. (2021). *BMJ Glob Health, 6(2)*, e004393, Availble at: https://gh.bmj.com/content/6/2/e004393.

Mosomi, J., & Wittenberg, M. (2020). The labor market in South Africa, 2000–2017: The legacy of apartheid and demand for skills have resulted in high, persistent inequality and high unemployment. *IZA World of Labor No. 475*, Available at: https://wol.iza.org/articles/the-labor-market-in-south-africa/long.

Motsoeneng, R. P. (2022). Public sanitation challenges in south african local government. *Administratio Publica, 30*(4), Available at: https://hdl.handle.net/10520/ejc-adminpub_v30_n4_a10.

Mubangizi, J. C., & Mubangizi, B. C. (2005). Poverty, human rights law and socio-economic realities in South Africa. *Deveplopment in South Africa, 22*, 277–290. Available at: https://doi.org/10.1080/03768350500163311.

Munslow, B., & Fitzgerald, P. (1994). South Africa: the sustainable development challenge. *Third World Quarterly, 15*(2), 227–242. Available at: https://www.jstor.org/stable/3993144.

Nagar, M. (2023). South Africa is failing to live up to its constitution. Gains made since democracy are being squandered—report. *The Converstion*, pp. https://theconversation.com/south-africa-is-failing-to-live-up-to-its-constitution-gains-made-since-democracy-are-being-squandered-report-209165.

National Development Plan. (2012). *National Development Plan—2030.* Available at: https://www.gov.za/sites/default/files/Executive%20Summary-NDP%202030%20-%20Our%20future%20-%20make%20it%20work.pdf: Panning Commission, Republci of South Africa.

National Planning Commission. (2021). *Mapping the national development plan to the united nations and African union sustainable development agendas.* National Planning Commission. Republic of South Africa.

National Planning Commission. (2011). *Diagnostic Overview.* Available at: https://www.gov.za/sites/default/files/gcis_document/201409/npcdiagnosticoverview1.pdf: The Presedency. Republic of South Africa.

National Planning Commission. (2012). *Mapping the national development plan to the united nations and African Union sustainable development agendas.* Available at: https://www.nationalplanningcommission.org.za/assets/Documents/Mapping_the_National_Development_Plan_to_the_United_Nations_and_African_Union_Sustainable_Development_Agendas_17_June_2021.pdf: National Planning Commission. Republic of South Africa.

NDoH. (2016). *South Africa demographic and health survey 2016: Report.* Available at: https://dhsprogram.com/pubs/pdf/FR337/FR337.pdf: National Department of Health (NDoH), Statistics South Africa (Stats SA), South African Medical Research Council (SAMRC), and ICF.

Ngubane, M., & Pillay, S. (2023). An exploration of the localization of the sustainable development goals (SDGS) in municipalities: A case of south African municipalities. *Journal of Public Administration, 5*, 1–10. Available at: https://doi.org/10.22259/2642-8318.0501001.

Nordling, L. (2020). South Africa flattens its coronavirus curve—and considers how to ease restrictions. *Sciences*, Available at: https://www.science.org/content/article/south-africa-flattens-its-coronavirus-curve-and-considers-how-ease-restrictions.

Ntshavheni, K. (2023). Minister Khumbudzo Ntshavheni, the Minister in Presidency: Planning, Monitoring and Evaluation. In S. SA, *Sustainable Development Goals: Country Report 2023* (pp. 4–5). Department of Statistics. Republic of South Africa. Available at: https://www.statssa.gov.za/MDG/SDG_Country_report.pdf.

OECD. (2019). *Education at a Glance: OECD Indicators.* Paris. Available at: https://www.oecd.org/education/education-at-a-glance/EAG2019_CN_ZAF.pdf: OECD.

OECD. (2022). *Financing SMEs and Entrepreneurs.* Available at: https://doi.org/10.1787/e9073a0f-en: Organization for Economic Co-operation and Development.

OIS. (2000). *Integrated Sustainable Rural Development Strategy.* Available at: https://www.gov.za/documents/other/integrated-sustainable-rural-development-strategy-17-nov-2000: Republic of South Africa.

OIS. (2020). Minister Zweli Mkhize reports first case of Coronavirus Covid-19. *Official Information Services*, pp. Available at: ww.gov.za/news/media-statements/minister-zweli-mkhize-reports-first-case-coronavirus-covid-19–05-mar-2020.

OIS. (2022). *Public Works and Infrastructure on work opportunities created by Expanded Public Works Programme.* Available at: https://www.gov.za/news/media-statements/public-works-and-infrastructure-work-opportunities-created-expanded-public: Official Information Service. Republic of South Africa.

OIS. (2023). Speaker Nosiviwe Mapisa-Nqakula: Speakers' Forum Gender Summit. *Official Information and Services. Government of South Africa*, pp. Avaialble at:https://www.gov.za/news/speeches/speaker-nosiviwe-mapisa-nqakula-speakers-forum-gender-summit-01-aug-2023.

OIS SA. (2023). Minister Joe Phaahla: Universal Health Coverage in South Africa. *Official Information Services. Republic of South Africa*, pp. Available at: https://www.gov.za/news/speeches/minister-joe-phaahla-universal-health-coverage-south-africa-12-dec-2022.

Oosthuizen, M. (2018). *Recognition of unpaid care work has been advocated by SDG5.4. Unpaid work is a key element of economic activity and welfare of an individual. It includes caring of child and elder peoples, and household chores. A certain part of our day is consumed while.* University of Cape Town: Development Policy Research Unit. Available at: https://commerce.uct.ac.za/poverty-inequality/articles/2019-09-23-counting-womens-work#:~:text=South%20Africa's%20women%20carry%20unpaid%20care%20burden&text=9%20billion%20in%202010.,production%20was%20contr.

OXFAM. (2020). *Reclaiming Power: Womxn's Work and Income Inequality in South Africa.* Available at: https://www.oxfam.org.za/wp-content/uploads/2020/11/oxfam-sa-inequality-in-south-africa-report-2020.pdf: OXFAM South Africa.

Panchia, Y. (2021). The 1956 Women's March In South Africa Still Resonates Today. *Forbes Africa*, pp. Available at: https://www.forbesafrica.com/focus/2021/08/09/the-1956-womens-march-still-resonates-today/.

Patel, L. (2021). Social security and social development in South Africa. In A. Oqubay, F. Tregenna, & I. Valodia, *The Oxford Handbook of the South African Economy* (pp. 869–888). Oxford Academic. Available at: https://doi.org/10.1093/oxfordhb/9780192894199.013.38.

Patel, L., & Triegaardt, J. (2008). South Africa: Social security, poverty alleviation and development. In J. Midgley, & K.-l. Tang, *Social Security, the Economy and Development* (pp. 85–109). Palgrave Macmillan. Avaible at: https://doi.org/10.1057/9780230582194_4.

Patel, L., Khan, Z., & Englert, T. (2018). How might a national minimum wage affect the employment of youth in South Africa? *Development Southern Africa, 37*(1), Available at: https://doi.org/10.1080/0376835X.2018.1552556.

Rahman, M. A., Halder, H. R., Rahman, M. S., & Parvez, M. (2021). Poverty and childhood malnutrition: Evidence-based on a nationally representative survey of Bangladesh. *PLoS One, 16*(8), Available at: https://doi.org/10.1371/journal.pone.0256235.

Ramaphosa, C. (2023). His Excellency Mr Cyril Ramaphosa, President of The Republic Of South Africa. In S. SA, *Sustainable Development Goals: Country Report 2023* (pp. 6–7). Department of Statistics. Republic of South Africa. Avaialble at: /https://www.statssa.gov.za/MDG/SDG_Country_report.pdf.

Ranchhod, V., & Daniels, R. C. (2020). *Labour market dynamics in South Africa in the time of COVID-19: Evidence from Waves 1 and 2 of the NIDSCRAM survey.* National Income Dynamics Study (NIDS)—Coronavirus Rapid Mobile Survey (CRAM.

REDIS. (2024). *Renewable Energy Data and Information Service.* Available at: http://redis.energy.gov.za/power-production/: Department of Mineral Resource and Energy. Republic of South Africa.

Reuters. (2023). South Africa pursues universal health cover in face of headwinds. *Reuters*, pp. https://www.reuters.com/world/africa/south-africa-pursues-universal-health-cover-face-headwinds-2023-06-13/.

Reyes, H., Pérez-Cuevas, R., Sandoval, A., Castillo, R., Santos, J. I., Doubova, S. V., & Gutiérrez, G. (2004). The family as a determinant of stunting in children living in conditions of extreme poverty: a case-control study. *BMC Public Health, 4*, Article number: 57, Available at: https://doi.org/10.1186/1471-2458-4-57.

Royo, M. G., Diep, L., J. M., Mukanga, P., & Parikh, P. (2022). Linking the UN sustainable development goals and african agenda 2063: Understanding overlaps and gaps between the global goals and continental priorities for Africa. *World Development Sustainability*, Available athttps://doi.org/10.1016/j.wds.2022.100010

RSA. (1994, November 23). Notice no. 1954 of 1994 parliament of the republic of south africa white paper on reconstruction and development. *Government Gazzet*, Available at: https://www.gov.za/sites/default/files/governmentgazetteid16085.pdf.

Sachs, J. D., Kroll, C., Lafortune, G., Fuller, G., & Woelm, F. (2023). *Sustainable development report 2023.* University Printing House.

SADC. (2020). *SADC Regional Indicative Strategic Development Plan (RISDP) 2020–2030.* Gaborone, Available at: https://www.sadc.int/sites/default/files/2021-08/RISDP_2020-2030.pdf: Southern African Development Community (SADC) Secretariat.

SAGNA. (2022, October 26). Government to step up spending on social wage. *South African Government News Agency*, Available at: https://www.sanews.gov.za/south-africa/government-step-spending-social-wage.

SAPS. (2023). *Annual Crime Stats 2022/2023 Crime Stats Presentation.* Avaialbel at: https://www.saps.gov.za/services/crimestats.php: Crime Registrar Head Office, South African Police Service (SAPS).

Scully, B., & Moyo, T. (2021). *Critical Sociology, 48*(3), Available at: https://doi.org/10.1177/0896920521103556.

Seedat, S., & Rondon, M. (2021). Women's wellbeing and the burden of unpaid work. *BMJ374:n1972*, Available at: https://doi.org/10.1136/bmj.n1972.

Seekings, J., & Nattrass, N. (2020). Covid vs. Democracy: South Africa's Lockdown Misfire. *Journal of Democracy, 31*(4), Available at: https://muse.jhu.edu/article/766188.

Sibeko, B., & Isaacs, G. (2020). *A fiscal stimulus for South Africa.* Witwatersrand. https://iej.org.za/wp-content/uploads/2020/08/A-fiscal-stimulus-for-South-Africa-Final-IEJ.pdf: Institute for Economic Justice.

Simelane, S., Sibeko, S., & Lwazikazi. (2023). *Governmnet Understanding Ananlysis 2011/12–2020/21: Department of Higher Education and Training.* https://www.parliament.gov.za/storage/app/media/PBO/Budget_Analysis/2023/3-june/05-06-2023/Analysis_of_Governemnt_Underspending_Higher_Education_and_Training_22_May_2023.pdf: Department of Higher Education and Training, Republic of South Africa.

Sinai, A. (2014). Macroeconomic policy challenges and choices in a time of crises. *Challenges*, 53–93.https://doi.org/10.2753/0577-5132520405

Singh, U. K. (2023). Realization of Amrit Kaal through Sustainable Development Goals. https://www.financialexpress.com/business/defence-realization-of-amrit-kaal-through-sustainable-development-goals-2996437/.

Soudien, C. (2022). Foreword: Struggling with the Weight of 'Race' and Racism. In G. Houston, M. Kanyane, & Y. D. Davids, *Paradise Lost: Race and Racism in Post-Apartheid South Africa. Volume 28* (pp. 7–10). Brill. Available at: 9789004515949-front-1.pdf.

Spilimbergo, A., Schindler, M., & Symansky, S. A. (2009). *Fiscal multipliers.* International Monetary Fund. https://doi.org/10.5089/9781462372737.004.

STATS SA. (2016a). *Employment, unemployment, skills and economic growth.* Avaialble at: https://www.statssa.gov.za/presentation/Stats%20SA%20presentation%20on%20skills%20and%20unemployment_16%20September.pdf: Statistics of South Africa. Republic of South Africa.

STATS SA. (2016b). *Millennium development goals: Final coutry reprot South Africa.* Available at: https://www.statssa.gov.za/MDG/MDG_Country_Report_Final_Executive%20Summary_Feb2016.pdf: Department of Statistics. Republic of South Africa.

STATS SA. (2018). *Men, Women and CHildren: Finding of the living condition Survey 2014/15 (Report No. 03–10–02 (2014/15).* Avaialble at: https://www.statssa.gov.za/publications/Report-03-10-02%20/Report-03-10-02%202015.pdf: Statistics South Africa.

STATS SA. (2019a). *Inequality Trends in South Africa: A multidimensional diagnostic of inequality/ Statistics South Africa.* Available at: https://www.statssa.gov.za/publications/Report-03-10-19/Report-03-10-192017.pdf: Statistics South Africa. Republic of South Africa.

STATS SA. (2019b). *Sustainable Development Goals: Country report 2019.* Available at: https://www.statssa.gov.za/MDG/SDGs_Country_Report_2019_South_Africa.pdf: Department of Statistics. Republic of South Africa.

Stats SA. (2021). *GDP: Quantifying SA's economic performance in 2020.* Available at: https://www.statssa.gov.za/?p=14074: Department of Statistics. Republic of South Africa.

Stats SA. (2022a). *Gender Series Volume IX: Women Empowerment, 2017–2022.* Available at: https://www.statssa.gov.za/publications/Report-03-10-26/Report-03-10-262022.pdf: Statistics South Africa. Republic of South Africa.

STATS SA. (2022b). *South Africa's Income & Expenditure Survey Takes Off.* Available at: https://www.statssa.gov.za/?p=15911: Statistics of South Africa. Republic of South Africa.

Stats SA. (2023a). *Fact Sheet on Access to Tertiary Education Country Comparison using Gross Enrolment Ratio (2023).* Available at: https://psetresearchrepository.dhet.gov.za/post/768: Department of Higher Education and Training. Republic of South Africa.

STATS SA. (2023b). *General Household Survey 2022.* Available at: https://www.statssa.gov.za/publications/P0318/P03182022.pdf: STATISTICS SOUTH AFRICA.

STATS SA. (2023c). *Macro-Social Report- 2022.* Availabe at: https://mistra.org.za/wp-content/uploads/2023/05/Macro-Social-Report-2023-ONLINE-VERSION-24-MAY-compressed.pdf: Stats SA.

STATS SA. (2023d). National poverty lines. *Statistical Release*, Available at: https://www.statssa.gov.za/publications/P03101/P031012023.pdf.

STATS SA. (2023e). *Quarterly labour force survey quarter 3: 2023.* Available at: https://www.statssa.gov.za/publications/P0211/P02113rdQuarter2023.pdf: Deaprtment of Statistic. Republic of South Africa.

STATS SA. (2023f). *Quarterly labour force survey (QLFS) Q3:2023.* Avaialble at: https://www.statssa.gov.za/publications/P0211/Presentation%20QLFS%20Q3%202023.pdf: Department of Statistics. Republic of South Africa.

STATS SA. (2023g). *Quarterly Labour Force Survey: Quarter 3: 2023.* Available at: https://www.statssa.gov.za/publications/P0211/P02113rdQuarter2023.pdf: Department of Statistics. Republic of South Africa.

Sulla, V., Zikhali, P. C., & Facundo, P. (2022). *Inequality in southern africa: an assessment of the southern african customs union.* Avaialable at: http://documents.worldbank.org/curated/en/099125303072236903/P1649270c02a1f06b0a3ae02e57eadd7a82: World Bank.

The Economist. (2019). High Unemployment is a symtom of South Africa's Economic Malaise. *The Economist*, pp. Available at: https://www.economist.com/special-report/2019/04/25/high-unemployment-is-a-symptom-of-south-africas-economic-malaise.

The Presidency. (2011). *Diagnostic report.* https://static.pmg.org.za/docs/110913NPCpp-edit_0.pdf: The Planning Commission,, Republic of South Africa.

UN. (2003). *Indicators for monitoring the Millennium Development Goals: Definitions, Rationale, Concepts and Sources.* Available at: http://mdgs.un.org/unsd/mdg/Resources/Attach/Indicators/HandbookEnglish.pdf: United Nations.

UN. (2014). *Statement delivered on the occasion of the 69tm session of the united nations general assembly general debate by president 3acob zuma of the republic of South Africa.* Available at: https://www.un.org/en/ga/69/meetings/gadebate/pdf/ZA_en.pdf: United Nations.

UN. (2015a). *69/269. Statute of the United Nations Nelson Rolihlahla Mandela Prize.* Avaialble at: https://daccess-ods.un.org/tmp/5390560.03093719.html: United Nations.

UN. (2015b). *Resolution adopted by the General Assembly on 2 April 2015.* Available at: https://documents-dds-ny.un.org/doc/UNDOC/GEN/N15/095/23/PDF/N1509523.pdf?OpenElement: United Nations.

UN. (2015c). *Resolution adopted by the General Assembly on 25 September 2015.* Available at: https://www.un.org/en/development/desa/population/migration/generalassembly/docs/globalcompact/A_RES_70_1_E.pdf: United Nations.

UN. (2016). *Department of Economic and Social Affairs: Sustaiable Development.* https://sdgs.un.org/goals/goal3#targets_and_indicators: United Nations.

UN. (2017). *Agenda 2063: The Africa We Want.* Available at: https://sdgs.un.org/publications/sustainable-development-goals-sdgs-and-africas-agenda-2063-18059: UNited Nations.

UN. (2019). *SOUTH AFRICA'S VOLUNTARY NATIONAL REVIEW (VNR) REPORT 2019.* Available at: https://sustainabledevelopment.un.org/content/documents/23402RSA_Voluntary_National_Review_Report__9_July_2019.pdf: United Nations.

UN. (2023a). *A Call to action for a better world: Nelson Mandela.* Available at: https://southafrica.un.org/sites/default/files/2023-07/NMF%20UNIC%20WEB.pdf: United Nations South Africa.

UN. (2023b). *ENTREPRENEURSHIP STRATEGY REVIEW: South Africa.* Avaialble at: https://unctad.org/system/files/official-document/tcsdiaeinf2023d3_en.pdf: United Nations Conference on Trade and Development.

UN. (2023c). *Global Multidimensional Poverty Index (MPI) 2023.* Available at: https://hdr.undp.org/system/files/documents/hdp-document/2023mpireportenpdf.pdf: United Nations Development Programme and Oxford Poverty and Human Development.

UNESCO. (2023). *Building better formal TVET systems: principles and practice in low- and middle-income countries.* Avaialb;e at: https://unesdoc.unesco.org/ark:/48223/pf0000386135: United Nations Educational, Scientific and Cultural Organization.

UNICEF. (2020a). *A New Era for Girls: Taking stock of 25 years of progress.* Available at: https://data.unicef.org/resources/a-new-era-for-girls-taking-stock-of-25-years-of-progress/: UNICEF.

UNICEF. (2020b). *South Africa Nutrition Brief.* Available at: https://www.unicef.org/southafrica/reports/nutrition-brief-2020: United Nations International Children Emergemcy Fund.

UNICEF. (2023). *Maternal mortality declined by 34 per cent between 2000 and 2020.* Avaialble at: https://data.unicef.org/topic/maternal-health/maternal-mortality/: UNICEF.

UNICEF. (2024). *Cross-sector indicators: South Africa.* Available at: https://data.unicef.org/resources/data_explorer/unicef_f/?ag=UNICEF&df=GLOBAL_DATAFLOW&ver=1.0&dq=ZAF.WS_PPL_S-B.&startPeriod=1970&endPeriod=2024: United Nations International Children's Emergency Fund.

Valodia, I. (2023, September 14). South Africa can't crack the inequality curse. Why, and what can be done. *The Conversation*, pp. https://theconversation.com/south-africa-cant-crack-the-inequality-curse-why-and-what-can-be-done-213132.

Verick, S., & Islam, I. (2010). The Great Recession of 2008–2009: Causes, Consequences and Policy Responses. *IZA Discussion Paper No. 4934*, https://docs.iza.org/dp4934.pdf.

Verschaeve, J., Delputte, S., & Orbie, J. (2016). The rise of policy coherence for development: A multi-causal approach. *The European Journal of Development Research, 28*, 44–61. Available at: https://doi.org/10.1057/ejdr.2015.74.

WEF. (2023). *Fostering effective energy transition: 2023 edition.* Available at: https://www3.weforum.org/docs/WEF_Fostering_Effective_Energy_Transition_2023.pdf: World Economic Forum.

Wekwete, N. N. (2017). Gender and the care economy in Southern Africa: Policy implications. *Women's economic empowerment in the changing world of work* (pp. 1–5). UN Women. Available at: /https://www.unwomen.org/sites/default/files/Headquarters/Attachments/Sections/CSW/61/meetings/Naomi%20Wekwete%20-%20The%20care%20economy%20in%20Southern%20Africa.pdf.

Wet-Billings, N. D. (2023). Perpetuation of household food insecurity during COVID-19 in South Africa. *Journal of Health, Population and Nutrition, 42*, Article number 96, Available at: https://doi.org/10.1186/s41043-023-00441-y.

WHO. (2020). What you need to know about malnutrition & feeding children. *Blog post*, Availble at: https://www.unicef.org/southafrica/stories/what-you-need-know-about-malnutrition-feeding-children.

WHO. (2023). *Progress on Sanitation and Hygiene in Africa 2000–2022*. Available at: https://washdata.org/: World Helath Organizations.

Winkler, H. (2021). Why southern Africa's interior is an ideal place to generate solar energy. *The Conversation*, pp. Available at: https://theconversation.com/why-southern-africas-interior-is-an-ideal-place-to-generate-solar-energy-161030.

World Bank. (2018a). *Overcoming poverty and inequality in South Africa*. Available at: https://documents1.worldbank.org/curated/en/530481521735906534/pdf/124521-REV-OUO-South-Africa-Poverty-and-Inequality-Assessment-Report-2018-FINAL-WEB.pdf: World Bank.

World Bank. (2018b). *Overcoming poverty and inequality in South Africa: Overcoming poverty and inequality in South Africa*. Available at: https://documents1.worldbank.org/curated/en/530481521735906534/pdf/124521-REV-OUO-South-Africa-Poverty-and-Inequality-Assessment-Report-2018-FINAL-WEB.pdf: World Bank .

World Bank. (2021). *Social Assistance Programs and Systems Review*. Available at: https://documents1.worldbank.org/curated/en/238611633430611402/pdf/South-Africa-Social-Assistance-Programs-and-Systems-Review.pdf: World Bank.

World Bank. (2022a). *Correcting curse: Poverty and shared properity-2022*. World Bank.

World Bank. (2022b). Fact Sheet: An Adjustment to Global Poverty Lines. *FACTSHEET*, pp. available at: https://www.worldbank.org/en/news/factsheet/2022/05/02/fact-sheet-an-adjustment-to-global-poverty-lines#:~:text=As%20differences%20in%20price%20levels,%242.15%20per%20person%20per%20day.

World Bank. (2022c). *Inequality in southern africa: an assessment of the Southern African customs union*. The World Bank.

World Bank. (2022d). *Inequality in Southern Africa: An assessment of the southern african customs union*. Available at: ?https://documents1.worldbank.org/curated/en/099125303072236903/pdf/P1649270c02a1f06b0a3ae02e57eadd7a82.pdf: World Bank.

World Bank. (2023a). *Poverty & equity brief: South Africa*. Available at: https://pip.worldbank.org/country-profiles/ZAF: The World Bank.

World Bank. (2023b). *GDP per capita (current US$)—South Africa*. Available at: https://data.worldbank.org/indicator/NY.GDP.PCAP.CD?locations=ZA-ZG-CF-XQ. accessed on 20/12/2023: The World Bank.

World Bank. (2023c). Measuring Poverty. *Understanding Povert*, p. Available at: https://www.worldbank.org/en/topic/measuringpoverty.

World Bank. (2023d). *The world bank in South Africa*. Available at: https://www.worldbank.org/en/country/southafrica/overview#:~:text=South%20Africa%20remains%20a%20dual,coefficient%20of%200.67%20in%202018.: World Bank.

World Bank. (2024a). *Gender data portal: South Africa*. Available at: https://genderdata.worldbank.org/countries/south-africa/.

World Bank. (2024b). *Going beyond the infrastructure funding gap: A South African perspective*. Avaialble at: https://www.dbsa.org/sites/default/files/media/documents/2024-01/Going%20Beyond%20the%20Infrastructure%20Funding%20Gap%20Publication%20%28WB-RAS%29.pdf: World Bank.

Xu, Z., Chau, S. N., Chen, X., Zhang, J., Li, Y., Dietz, T., … Liu, J. (2017). Assessing progress towards sustainable development over space and time. *Nature, 577*, 74–78.

Zalk, N. (2014). Industrial policy in a harsh climate: The case of South Africa. In N. Zalk, *Transforming Economies: Making industrial policy work for growth, jobs and development—Introduction* (pp. 327–354). International Labour Organization. Available at: https://www.ilo.org/global/publications/books/WCMS_315677/lang--en/index.htm.

Ziady, H. (2023). South Africa's energy crisis deepens as blackouts hit 12 hours a day. *CNN*, pp. Available at: https://edition.cnn.com/2023/01/18/energy/ramaphosa-davos-south-africa-blackouts/index.html.

Chapter 4
Impact of UN's Sustainable Development Goals on Human Development in India and South Africa

Introduction

Amartya Sen in his work *Development as Freedom* noted that despite the unprecedented rise in global wealth many people are still deprived of elementary freedom (Sen, 2000). In answering 'What is development?' Slim (1999) stressed that real development is more important than economics and its measuring parameters. He further emphasized development for all societies rather than limiting the concept of development to a third-world problem. Development is the result of interaction between two different parties, and two different countries, and therefore development is dependent on the success of relationships.

Sen's idea of freedom focuses on the expansion of capabilities. His vision of development revolutionized the parameters for measuring the degree of development achieved by a country when he shifted the locus of measurement from economic growth to progress in human capabilities necessary for human development. Sen's approach to development has provided a new approach to understanding of poverty and development. As Mackie (2012) succinctly observed, Sen's approach to development has placed human-centric poverty reduction as core to global development agendas.

India entered the era of Liberalization, Privatization, and Globalization and on the other side, South Africa stepped into a new era with a new regime coming into power with a great political shift in the post-apartheid era. Both countries have used various tools to accelerate their economic growth, which is considered an engine of development to ensure that the profit of growth is shared by their population and both countries have adopted various measures to reduce poverty, hunger, unemployment, and inequality. However, the progress of India and South Africa are modest.

U. K. Singh, *Regional Dimensions of Human Development in India and South Africa*, Advances in Geographical and Environmental Sciences,
https://doi.org/10.1007/978-981-97-5846-3_4

Causality Between Economic Growth and Poverty Alleviation

Inclusive economic growth and development are closely linked with poverty reduction, whereas poverty alleviation is core to the principle of sustainable development (Zhu et al., 2022). In what has become an oft-cited maxim, poverty in any form is a serious threat to humanity (Park & Mercado, 2018; Omar & Inaba, 2020). Global agenda-2030 has identified poverty as a root cause of all forms of deprivation. Thus, SDG1 (No Poverty) is recognized as *numero-uno* to achieve other goals of SDGs.

Poverty is a multifaceted phenomenon. The poor have many things in common, not income alone (World Bank, 1989). Poverty goes far beyond the notion of monetary estimation and is much more than extremely low income or insufficient dietary intake to support a healthy life (Ananad & Sen, 1997; Siddiqui et al., 2020). Bourguignon and Chakravarty (2003) questioned the understanding of poverty solely on monetary variables. Similarly, Anand et al. (2020) called for the inclusion of non-monetary dimensions in the estimation of poverty. Existing studies indicate that poverty's scourge is identified by inadequate access to education and healthcare, prone to diseases, poor quality of housing, unsafe surroundings, social exclusion, and extremely low participation in decision-making (Rolfe et al., 2020; Sen, 2011; UN, 1995). From the lens of human development, poverty is the limitation on an individual in making choices and denial of opportunities for a reasonable life (Singh, 2022; Anand & Ravallion, 1993; UN, 1990).

Regardless of any approach of measuring poverty at the national or regional level, resolving the problem of destitution of an individual rely on command and guidance of multiple system and policies (Huang et al., 2023). The situation of poor and poverty can be solved by efficient and effective policy interventions (Salecker et al., 2020). According to United Nations statistics, across 110 countries nearly 1.1 billion people live in multidimensional poverty (UNDP, 2023a, 2023b). An estimated 702 and 828 million people starved in 2021 and COVID-19 has pushed an additional 150 million to hunger (FAO, IFAD, UNICEF, WFP, WHO, 2022). World Bank estimates that an approximate 674 million people live in extreme poverty (World Bank, 2022).

The ambitious goal to end poverty is rooted in the philosophy of human development. SDGs serve as development's destination. The accomplishment of SDGs depends on the localization of Goals. Every country possesses a unique set of strength and challenges at both national and regional levels. Global agenda provides the opportunity to its member countries to identify their challenges, build the targets and indicators according to local needs, converge the local targets with Global Agenda-2030 (Janus & Keijzer, 2013). The interaction between SDGs and the development strategies of host countries amplifies the fulfillment of goals with accountability.

World leaders regard containing poverty as a monumental task and a key to human development. They joined hands together and made a historic agreement to end poverty and ensure the well-being of every individual through the Sustainable Development Goals (SDGs) (UN, 2015). The multiple dimensions of poverty are encapsulated in SDGs. A set of 17 goals with 169 targets are interlinked and balance the social, economic, and environmental dimensions of sustainable development, to be achieved by 2030. SDGs are an expansion of Millennium Development Goals (MDGs), to finish the incomplete task through comprehensive and relevant indicators (Singh, 2022; Blanc, 2015). SDGs provide a way forward to a humane world.

Background

India and South Africa both have an untapped potential in the form of the large proportion of youth population. India and South Africa are in the position of demographic dividend. Both the countries are experiencing youth bulge. A quarter of India's population are below the age of 14 years, whereas in South Africa the proportion is 28%. The size of the working-age population (1–64 years) in India is 68% whereas the corresponding figure for South Africa is 66%.

Studies have shown that the sole dependence on rising economic growth is not sufficient for poverty reduction (Fajnzylber, 2018). Any mismatch between growth and development has undesirable consequence in the long term and also negative effect on poverty, hunger, employment and other wellbeing and macro indicators (Cerra et al., 2021). To the contrary, inclusive growth is key to ensure upliftment of people from deprivation. Rodrik (2000) even argued that it may possible for a country to reduce its poverty through redistributive measure without any growth in economy. It could be possible though focused policy intervention that leads to rise in poor's income. For example, investment in education and public health has capacity to enhance the income of an individual as well as overall economy.

The accomplishment of SDGs depends on policy design for human development. Development trajectory of India and South Africa should be vitally linked with investment in public health and education. Investment in human capital and inclusive growth is seminal for achieving SDGs. India and South Africa recognizes the youth potential as a key to achieve SDGs. To reap the benefit of the demographic dividend, it is important to transform quantity (large youth population) in to quality (human capital). India and South Africa have made every effort to improve the outcomes in education and ensure access of health care to all. It is presumed that ending poverty is dependent of strategies implemented to improve the health and education of an individual. India and South Africa has set eyes on long term development goals. India has envisaged *Amrit Kaal* (pious period) and South Africa has aligned her development targets with the National Development Plan (NDP)-2030 with the target year 2063. The route of *Amrit Kaal* and NDP-2030 goes thorough the SDGs. Both countries have committed to creating an empowered and inclusive society with a sustainable and economic growth model in tuned with SDGs to realize human development, and other objectives as well (PIB, 2023). However, with unprecedented global shock India and South Africa grappled with other countries in the uncertainty caused by the COVID-19 pandemic and the Ukraine-Russia war.

To redeem the idea of *Amrit Kaal* and NDP-2030, there is a need to give extra thrust to SDGs. For advancement in the social indicators, the economy needs strong economic and environment indicators. To address this concern, what strategy, acknowledging the synergies and tradeoffs between the SDGs, should India and South Africa pursue to meet their critical challenges of creating a thriving and sustainable economy? The purpose of this study is to identify and quantify the impact of such SDGs on the GDP in India and South Africa. The study also seeks to figure out the strategy for India to minimize tradeoffs and maximize synergy between the SDGs.

This chapter identified the particular SDGs and their corresponding indicators based on inclusive and sustainable growth agenda of India and South Africa. To address the challenges and opportunities, world's development partners such as United Nations, World Bank, and IMF engaged with the governments at national and subnational level in India and South Africa (Worl Bank, 2019; Benedek et al., 2021). Vision Amrit kaal aims to creates 90 million jobs by 2030 and attaining 8% GDP growth each year in succeeding decades (McKinsey & Company, 2023), whereas South Africa is determined to create 11 million employment through NDP-2030 (NPC SA, 2011). The development agenda of both the countries is dependent on their capacity to unleash the potential of manpower by 2030. The realization of vision necessitates the identification and quantification of the impact of SDGs that can lead to achievement of Amrit kaal (Singh, 2023). Determination of public policy is indispensable to accomplish human development. The Policy's effectiveness depends on understanding poverty and its cause as well as knowledge of who the poor are and where they live.

The effective implementation of the SDGs is dependent on locating relevant targets and suitable indicators, while goals determine what strategy is needed for sustainable development in decentralize manner, in accordance with local area. India and South Africa has recognized that the youth potential and mainstreaming women and marginalized section is vital to achieve robust economic growth. In this sustainable and inclusive growth model, knowledge creation and innovations are key determinant in further advancement of economic and social progress simultaneously. However, maintaining balance between growth and sustainability is always a challenge for policymakers, especially in developing countries like India and South Africa. The interdependence between SDGs indicates the existence of synergies but also points toward tradeoffs (Pradhan et al., 2017). For some SDGs, interactions are strong, while others are blurred. For example, containing inequality across income groups and its impact on the environment could be negative or positive (Rao & Min, 2018). Most developing countries deal with complex socio-economic issues along with ecological sustainability simultaneously, posing a threat to the successful implementation of SDGs.

This chapter explores the causality between SDGs and GDP growth as proxy for India and South Africa during 2000–2022. It has identified targets of human development and their respective indicators in relation to India's focus on inclusive development under its Vision 2030 and South Africa's NDP 2030. It seeks to assess the impact of human development indicator on GDP growth rate and takes social goals of SDGs such as no poverty, zero hunger, health and wellbeing, quality education, gender parity, inequality and decent jobs for evaluation. Using multiple regression analysis to predict GDP growth of India through quantifying the impact of each variable, the result shows that health and education have significant impact on GDP growth rate. This chapter attempts to answer the following research questions:

RQ1: What is the impact of SDGs on economic growth in India and South Africa.

RQ2: What is the impact of SDGs on the human development in India and South Africa?

Literature Review

The SDGs are the blueprint to shift the world on sustainable path, a step toward better future for all, to be achieved by 2030 (UN, 2023). Unlike preceding development agenda that focuses more on economic growth, SDGs upholds equity, inclusion, and universality (Cerf, 2019). The SDGs are conceived as "integrated and indivisible" by design. SDGs are interlinked and interdependent (Tosun & Leininger, 2017) and should be implement as such (Luttikhuis & Wiebe, 2023). Existing scholarships have shown that correlation between SDGs is inevitable due to the comprehensiveness of goals (Scherer et al., 2018). These studies held that pursuing specific goal have spill over impact on others goals, which has been identified as a major drawback in implementing SDGs. Interactions likely range from indivisibility (progress in an SDG is contingent on progress of another) to cancellation (success of an SDG retard progress of another; Nilsson et al., 2016).

Therefore, each and every SDG act as input to the collective SDG framework; progress in one goal ripples to advancement of other SDGs. The indivisibility among SDGs is wide spread as axiomatic, and reinforcing evaluation of 'policy synergy' leads to identifying where tradeoff would be expensive (ISC, 2017). In practice, identifying policy coherence between and within the SDGs is a huge challenge but also a necessary action for policy makers.

SDGs should be pursued in asymmetric manner. Governments are free to choose the goals as per their own ambition in implementing the SDGs. For example, SDG 8 calls for increase in GDP continually which is in conflict with SDG 13 that aims to reduce carbon footprints. This conflict cause government incentive to contain greenhouse gas emission, however, these harmful activities are positively associated with GDP (Nature, 2023; Badulescu et al., 2019). Therefore, member states choose their goals as per local aspiration with more emphasis on indicators of human development that accelerate the economic activity consequently raising the GDP.

The primary challenge for policy makers is to synchronize the success of SDGs. To resolve this situation, researchers find solution in synergy and tradeoffs between the multiple policy objectives (Zhang et al., 2022; Dissanayake & Mahadev, 2020). Study of Pradhan et al. (2017a, 2017b, 2017c) is among the first to quantify the synergy and tradeoffs within and across the SDGs. They observed that SDG 1 (no poverty) has synergetic association with almost all goals (Pradhan P., Costa, Rybski, Lucht, & Kropp, A Systematic Study of Sustainable Development Goal (SDG) Interactions, 2017). Similar pattern is outlined in the study of Lusseau and Mancini (2019) who contends that SDG 10 (reduce inequality), SDG 12 (responsible consumption) and SDG 13 (climate change) are major obstacle to achieving SDGs across countries. While SDG1 (no poverty) and SDG 10 (reduce inequality) will have multiple contingent reinforcement to other SDGs. McGowan et al., (2019) observed the uneven connection between the goals, for instance the difficulty of integrating SDG 5 (gender equality) and SDG 16 (peace and governance). Zhao et al., (2022) noted that SDGs solution lies in the reclassification of status and vison, synchronizing multitudinous policies and resources, and accelerating and assimilating economic, technology & innovation, and political alliance. The COVID-19 pandemic has reoriented the development debate towards environmental priorities SDG 13 (Life on Land); SDG 14 (life under sea) and its positive correlation with SDG 3 (health and wellbeing) (Smith et al. 2021).

In a more recent study, Cling and Delecourt (2022) measured the correlation between the indicators of SDGs through the Multiple Factor Analysis. Their their study showed the strong link beween successful development strategies and improvement in national productive capacilty in tandem with progess in other dimensions of human development. Ugnich et al., (2021) study revealed that BRICS nations have high potential for the enhancement of human capital, however, it is always underperformed, which is mirrored in the implementation of policies to accompolish SDGs. It further reinforced the argument that in the fast changing world, the idea of sustaianble development of human capital is key to policy design for SDGs. It is built on the balanced approch to multiple aspects of human life, collectively, human development.

Role of Human Development's Indicators in Achieving SDGs

United Nations Development Programme theorized the human-centric approach in its first Human Development Report (HDR) finding expression in its famous maxim '*humans are the real wealth of nation*' (UN, 1990). The turn toward human development was a pivotal shift in the accounting of economic growth and development when the focus in accounting shifted to 'human capital' from the conventional physical capital. Human capital is the source of future earning, (Schultz, 1972).

Health, education and economy are directly linked to poverty. The three standard dimensions are dedicated to examine the progress attained by a person to sustain the kind of life they value, including longevity (Health), Knowledge (Education) and command over resources (Income). Health and knowledge are considered as 'human capital' and are required to expand human capability, while command over resources is a means to the other ends. Per person economic growth Gross National Income (GNI) is taken as proxy for command over resources since it provides access to health and education and other capabilities.

In respect to SDGs, human capital has the potential to improve the standard of living and environment in the long run. The quality of human capital along with physical capital is vital to realize optimal production efficiency (Payab et al., 2023).

The success of SDGs and inclusive growth is directly linked to investment in human capital (OECD & UNDP, 2019).

Health (SDG3) have the potential to advance other SDGs (Smith et al., 2023). Health is core to human capabilities, good health leverage human capacity, income and gender equity across the societies to achieve human development. Consumption of health care by an individual improve its stock of health, which further reciprocate into productive resources (Grossman, 1972). The World Health Organization (WHO) literature has recognized the interlinkages between health and economic growth (Sachs, 2001). Sen (2002) also identified health as a key element of social justice. The social elements of health straddles humanity, education (attainment and output), healthcare (access, attainment, and quality), the economy (sustainable growth) and to some extent create opportunities for decent employment which in return improve living condition, which promotes individual's wellbeing and standard of living (Cerf, 2023).

Education helps in acquiring skill set that improve chances of employability and better income and serve as protection to people living in abject poverty. Education is a key determinant in enhancing people's capacity and public awareness. Ho and et al. (2023) argued that education is key to capacity building and public awareness for global workforce, it should be developed and delivered through institutions. UNESCO literature revealed that 60 million people could exit poverty by adding two more year of schooling (UNESCO, 2023).

Hypothesis Development

The above discussed literature suggests that human capital have a positive impact on the GDP growth. It also supports the idea that poverty reduction has causal effect on other dimensions of human development as well as goals of SDGs. GDP growth has inverse relation with poverty. In India and South Africa, growth is significantly related to enrolment in higher education, decent jobs and prevalence of malnutrition among children.

Based on the literature this study analyzes eight indicators related to human development indicators across eight targets for six goals. These eight indicators are linked to multiple dimensions of human development. Table 4.3 represent the association between goals, targets, indicators and variables. Based on the aforesaid discussion, I have designed the following hypothesis with set of assertions.

Hypothesis: Implementation of SDGs will lead to realization of human development in India and South Africa.

Hypothesis

1. There is a direct relationship between Hunger and the GDP growth rate in India and South Africa.
2. There is a direct relationship between Health and the GDP growth rate in India and South Africa.

3. There is a direct relationship between Education and the GDP growth rate in India and South Africa.
4. There is a direct relationship between Decent Job and the GDP growth rate in India and South Africa.
5. There is an inverse relationship between Hunger and the GDP growth rate in India and South Africa.
6. There is an inverse relationship between Health and the GDP growth rate in India and South Africa.
7. There is an inverse relationship between education & gender parity and the GDP growth rate in India.

Data and Methodology

This study aims to investigate the corelation between human development indicators and India's economic growth denoted by GDP growth rate (annual %). Based on India and South Africa's long-term vison as envisaged in Amrit Kall and Agenda 2063 respectively, SDGs indictors in respect to human development were selected as per their availability for both India and South Africa. Panel annual data from 2001 to 2020 were used and their impact on the economic growth of both the countries was evaluated.

This study recognises the limitations of GDP in measuring overall wellbeing of a country. Therefore, this study used GDP growth rate as a proxy for economic growth. The annual GDP growth rate is taken as dependent variable and the data was obtained from World Bank indicators. The independent variables were selected on the basis of their accessibility and quantifiability. Subsequently, time series data of eight indicators related to human development and SDGs were identified and used for analysis.

The choice of indicators was done in accordance with its relevance and significance to the accomplishment of Amrit kaal and NDP-2030—pursuit of inclusive and developed economy. Data set were obtained from World Bank Indicators, United Nations Educational, Scientific and Cultural Organization (UNESCO) Institutes for Statistics, World Health Organizations (WHO), Ministry of Education, Government of India, Food and Agriculture Organizations (FAO) and International Labour Organizations (ILO). Consequently, aforementioned sources were used to navigate data in respect to specific indicators of SDGs and human development of India. Table 4.3 demonstrate the independent variables, their description, measurement and source and their connection with particular SDGs.

Empirical Model

The study set out to evaluate the following empirical model in equation to obtain the effectiveness of particular SDGs on the GDP growth rate of India and South Africa (Tables 4.1, 4.2 and 4.3).

Model for India

$$For\ India\ GDPIN_t = \beta_0 + \beta_1 HUNIN_t + \beta_2 LEBIN_t + \beta_3 MMRIN_t + \beta_4 NMRIN_t + \beta_5 U5MRIN_t + \beta_6 GERTUIN_t + \beta_7 GPITSIN_t + \beta_8 DJIN_t e_t$$

Model for South Africa

Table 4.1 Explanation of variables used in model for India

GDPIN	GDP growth Rate in Indi over a time period t in India
HUNIN	Prevalence of Hunger (Undernourishment % of total Population) in India
LEBIN	Life Expectancy at Birth in India
MMRIN	Maternal Mortality Rate in India
NMRIN	Neonatal Mortality Rate in India
U5MRIN	Under five Mortality Rate in India
GERTUEIN	Gross Enrolment Ratio (total) at upper secondary level (age group 18–22 years) in India
GPITSIN	Gender Parity Index at tertiary school enrollment (gross) (age group 18–22 years) in India
DJIN	Decent Job in India
e_t	Error term

Table 4.2 Explanation of variables used in model for South Africa

GDPSA	GDP growth Rate in Indi over a time period t in South Africa
HUNSA	Prevalence of Hunger (Undernourishment % of total Population) in South Africa
LEBSA	Life Expectancy at Birth in South Africa
MMRSA	Maternal Mortality Rate in South Africa
NMRSA	Neonatal Mortality Rate in South Africa
U5MRSA	Under five Mortality Rate in South Africa
GERTUESA	Gross Enrolment Ratio (total) at upper secondary level (age group 18–22 years) in South Africa
GPITSSA	Gender Parity Index at tertiary school enrollment (gross) (age group 18–22 years) in South Africa
DJSA	Decent Job in South Africa
e_t	Error term

Table 4.3 Variable explanation and measurement

Explanatory Variable	Variable Description	Source	Measurement	Related SDG
Hunger	Prevalence of Undernourishment	World Bank (2023a, 2023b, 2023c, 2023d, 2023e, 2023f, 2023g)	Proportion of undernourishment people in total population of India	2
Poverty, Hunger and Health	Life Expectancy at birth	World Bank (2023a, 2023b, 2023c, 2023d, 2023e, 2023f, 2023g); UNPA (2023)	Life Expectancy at birth	1.2 and 3
Health	Maternal Mortality Rate	World Bank (2023a, 2023b, 2023c, 2023d, 2023e, 2023f, 2023g); UNICEF (2023)	Death of women per 100,000 live births, due to any pregnancy and delivery related issues	3.1
Health	Neo Natal Mortality (death per 1000 births)	World Bank (2023a, 2023b, 2023c, 2023d, 2023e, 2023f, 2023g)	Death of children Under one year of age (death per 1000 births)	3.2
Health	Under Five Mortality Rate	World Bank (2023a, 2023b, 2023c, 2023d, 2023e, 2023f, 2023g)	Death of children under five year of age (death per 1000 births)	3.2
Education & Gender Parity	Gross Parity Index of School enrollment at tertiary (gross) level	World Bank (2023a, 2023b, 2023c, 2023d, 2023e, 2023f, 2023g); STATSA (2015); (Ministry of Education, 2023) Ministry of Education (2023)	Gender Parity Index at tertiary level (age group 18–22 years)	4 and 5
Education & Gender Parity	Gross Enrolment Ration in upper secondary	World Bank (2023a, 2023b, 2023c, 2023d, 2023e, 2023f, 2023g); (UIS STAT (2023); STATSA (2015); (Ministry of Education, 2023)	Gross Enrolment in Upper secondary (age group 18–22 years)	4 and 5
Decent Job	Decent job to all working population	World Bank (2023a, 2023b, 2023c, 2023d, 2023e, 2023f, 2023g)	Decent employment as share of total employment, based on ILO guidelines. The share of decent is determined by excluding vulnerable employment from total employment	8

$$
\begin{aligned}
GDPSA_t = & \beta_0 + \beta_1 HUNSA_t + \beta_2 LEBSA_t + \beta_3 MMRSA_t + \beta_4 NMRSA_t \\
& + \beta_5 U5MRSA_t + \beta_6 GERTUSA_t + \beta_7 GPITSSA_t + \beta_8 DJSA_t e_t
\end{aligned}
$$

Method

The Ordinary Least Square (OLS) regression Model (OLS) employed as technique to predict or understand the relation between the SDGs and GDP growth in India. OLS provides a specific route to measure "accuracy" and therefore provide an opportunity to choose "the best" constant from the given set of training data (the data we use to evaluate the accuracy). Further, the OLS regression technique lessens the sum of the squared differences between the actual value and predicted values for the dependent variables. The objective of the study finds its consistency with OLS regression, which is to identify the impact of numerous indicators of SDGs on GDP growth rate of India in pursuit of inclusive development.

Result

Tables 4.4 and 4.5 demonstrate the statistics of the variables employed for multiple regression analysis within the given time frame of this study for India and South Africa respectively. For India, statistics indicates that in an ideal condition the GDP growth rate is at 5.97%, prevalence of undernourishment at 16.4, life expectancy at birth at 67.29, MMR at 200.75, NMR at 31.37, U5MR at 57.86, GER in upper secondary (for both sexes) at 50.77, GPI tertiary education at 0.83, and decent jobs at 20.34. The coefficient of variation (CV) measures the data dispersion relative to the mean. The life expectancy's CV at 3.7 reflects least volatile among the variables, while GDP's CV at 54.29 appears as highly volatile.

The corresponding figure for South Africa is average GDP growth rate at 2.161%, prevalence of undernourishment at 5.1, life expectancy at birth at 59.9, MMR at 171.3, NMR at 12.835, U5MR at 54.65, GER in upper secondary (for both sexes) at 90.237, GPI tertiary education at1.282, and decent jobs at 89.705. The decent job's CV at 0.72 reflects least volatile among the variables, while GDP's CV at 121.84 appears as highly volatile.

The coefficient variance of GDP in both the countries appears to be highly volatile. The data substantiate the qualitative analysis in Chap. 3 and this chapter, which shows that India and South Africa experienced steeper decline in GDP growth rate during Covid-19. It also complements the data of life expectancy in India which show mere decline in recent years. The decent jobs in South Africa are least volatile because they are highly formalized.

Table 4.4 Statistical summary of all variables used for India

India	Mean	Minimum	Maximum	Standard Deviation	Coefficient Variation	VIF
Constant	5.97	−5.83	8.49	3.24105	54.29	
Prevalence of Undernourishment_	16.4	13	22	2.835861	17.29	16.043
Life expectancy at birth	67.29	63.09	70.91	2.550264	3.79	341.186
MMR_IN	200.75	103	371	83.1219	41.41	239.314
NMR_IN	31.37	20.2	43.3	7.050353	22.47	5185.217
U5MR_IN	57.86	32.4	88	17.66535	30.53	9050.198
GER (%) Both sexes upper secondary_ IND	50.77	32.8	67.07156	11.46314	22.58	140.605
GPI tertiary education	0.83	0.66	1.1	0.162152	19.54	112.829
Decent jobs_IND	20.34	16.41	26.84	3.731205	18.34	302.812

Table 4.5 Statistical summary of all variables used for South Africa

South Africa	Mean	Minimum	Maximum	Standard Deviation	Coefficient Variation	VIF
Constant	2.161	−5.96	5.6	2.633	121.84	
Prevalence of Undernourishment_ SA	5.1	3	8	2.835	55.59	8.801
Life Expectancy at birth	59.9	54	66	4.711	7.86	104.336
MMR	171.3	27	232	51.559	30.10	5.196
NMR_SA	12.835	11	16.9	2.03	15.82	23.102
U5MR_IN	54.65	33.7	79.2	18.585	34.01	137.464
GER (%), both sexes upper secondary_SA	90.273	71.241	111.663	15.935	17.65	15.35
GPI tertiary education	1.282	1.16	1.39	0.072	5.62	9.071
Decent jobs_SA	89.705	88.62	91.18	0.645	0.72	2.957

Tables 4.6 and 4.7 shows the Pearson correlation among the explanatory variables for India and South Africa. It helps in measuring the issue of multicollinearity. The issue of multipolarity emerges in a regression model when two or more highly independent variables are assessed (Farrar & Glauber, 1967; Aiken et al., 1991; Lukman et al., 2023). The adversity due to multicollinearity in regression model has been documented by several scholars (Mason, 1987; Vatcheva et al., 2016). Studies

show that the biased and unstable standard errors tend to unstable p-value, which is significant in examining the independent variables (Knijnenburg et al., 2009). This may lead to superficial data interpretation. The correlation analysis produced a combination of positive and negative output, however, the correlation between the indicators is strong in case of India while South Africa shows mix result.

The Pearson correlation between the variable used for India, as shown in Table 4.6, indicate strong correlation between the indicators. For example, prevalence of undernourishment and life expectancy at birth (−0.861); prevalence of undernourishment and MMR (0.870); prevalence of undernourishment and NMR (0.836); prevalence of undernourishment and U5MR (0.856); prevalence of undernourishment and GER at upper secondary (−0.851); prevalence of undernourishment and GPI tertiary education (−0.775); and prevalence of undernourishment and decent jobs (−0.772). Life expectancy at birth is highly correlated with all the health indicators, so as for GER and GPI in education show perfect correlation.

In contrary, Person correlation data for South Africa (see table 4.7), shows life expectancy in South Africa increase with undernourishment, the correlation between undernourishment and life expectancy is 0.920. It further demonstrates that the incidence of MMR, NMR and U5MR is improving with rise in undernourishment while the GPI at tertiary is decreasing.

Unit Root Test

One of the salient features of time series data is stationarity. This characteristic makes time series highly volatile, more often it is categorized by heteroskedasticity (Livieris et al., 2020). The data stationarity has significant impact on the efficiency of time series model (Livieris et al., 2021). The unit root test can be employed to identify differential frequency of time series data. This study conducts a unit root test on the data before running a multiple regression analysis to assess the impact of various SDGs on the GDP growth rate of India. It is an attempt to test the stationarity of given time series data.

The test provides the base to reject the null hypothesis id statistics < critical value and p value <0.05. This shows the non-exitance of time-dependent structure in given data. Further, it predicts that stationarity of given data based on the non-existence of unit root. The unit root test is depicted in Tables 4.8 and 4.9. Kwiatkowski-Phillips-Scmidt-Shin (KPSS) value indicate the presence/non-presence of a unit root in selected variables.

The result show no-presence in the series of under undernourishment, MMR, NMR, U5MR in both countries, with addition of decent jobs data series for South Africa.

The R^2 of utilized model quantifies extent of predication of dependent variables applying independent variables, regarding variance proportion (Wright, 1921; Chicco et al., 2021). The regression output demonstrates that the regression line is perfectly compatible with data. As shown in Tables 4.10 and 4.11, the employed model has high explanatory power with R^2 0.895 for India and 0.755 for South Africa. It also indicates that the variables used in model have influence on GDP growth rate. However, some variable shows moderate or insignificant association.

Table 4.6 Correlation between variables used for India

India	GDP_IND	Prevalence of Undernourishment_ India	Life expectancy at birth	MMR_IN	NMR_IN	U5MR_IN	GER (%) Both sexes upper secondary_IND	GPI tertiary education	Decent jobs_IN
GDP_IND	1.000								
Prevalence of Undernourishment_ India	0.098	1.000							
Life expectancy at birth	−0.203	−0.861	1.000						
MMR_IN	0.192	0.870	−0.971	1.000					
NMR_IN	0.305	0.836	−0.991	0.976	1.000				
U5MR_IN	0.272	0.856	−0.993	0.984	0.999	1.000			
GER (%) Both sexes upper secondary_ IND	−0.268	−0.851	0.987	−0.972	−0.987	−0.989	1.000		
GPI tertiary education	−0.364	−0.775	0.945	−0.869	−0.940	−0.933	0.948	1.000	
Decent jobs_IND	−0.332	−0.772	0.958	−0.877	−0.951	−0.943	0.950	0.993	1.000

Table 4.7 Correlation between variables used for South Africa

South Africa	GDP_SA	Prevalence of Undernourishment_ SA	Life Expectancy at birth	MMR	NMR_SA	U5MR_IN	GER (%) Both sexes upper secondary_SA	GPI tertiary education	Decent jobs_SA
GDP_SA	1.000								
Prevalence of Undernourishment_ SA	−0.722	1.000							
Life Expectancy at birth	−0.632	0.920	1.000						
MMR	0.516	−0.754	−0.832	1.000					
NMR_SA	0.520	−0.810	−0.851	0.585	1.000				
U5MR_IN	0.652	−0.919	−0.979	0.756	0.916	1.000			
GER (%) Both sexes upper secondary_SA	−0.614	0.879	0.923	−0.819	−0.836	−0.890	1.000		
GPI tertiary education	−0.603	0.736	0.748	−0.420	−0.896	−0.828	0.677	1.000	
Decent jobs_SA	−0.276	0.530	0.607	−0.454	−0.682	−0.658	0.609	0.458	1.000

Table 4.8 Unit root test for stationary data for India

Variables	kpss value	p_value	Unit Root
GDP_IND	0.27	0.10	Yes
Prevalence of Undernourishment_India	0.66	0.02	No
Life expectancy at birth	0.28	0.10	Yes
MMR_IN	0.80	0.01	No
NMR_IN	0.73	0.01	No
U5MR_IN	0.79	0.01	No
GER (%) Both sexes upper secondary_IND	0.16	0.10	Yes
GPI tertiary education	0.15	0.10	Yes
Decent jobs_IND	0.12	0.10	Yes

Table 4.9 Unit root test for stationary data for South Africa

Variables	kpss value	p_value	Unit Root
GDP_SA	0.59	0.02	No
Prevalence of Undernourishment_SA	0.15	0.10	Yes
Life Expectancy at birth	0.21	0.10	Yes
MMR	0.70	0.01	No
NMR_SA	0.64	0.02	No
U5MR_IN	0.76	0.01	No
GER (%) Both sexes upper secondary_SA	0.13	0.10	Yes
GPI tertiary education	0.28	0.10	Yes
Decent jobs_SA	0.36	0.09	No

Table 4.10 Multiple regression result for India

India	Multiple regression Result SDG Variable to growth rate		
	Coefficients	Standard Error	t Statistic
Intercept	−546.539	143.470	−3.809
Prevalence of Undernourishment IN	1.029	0.445	2.307
Life expectancy at birth	7.007	2.286	3.064
MMR_IN	0.022	0.058	0.383
NMR_IN	7.552	3.224	2.342
U5MR_IN	−2.547	1.700	−1.498
GER total (%) upper secondary _IND	−0.338	0.326	−1.036
GPI tertiary education	−11.718	20.681	−0.566
Decent jobs IN	−0.143	1.472	−0.097
R^2	0.895		
F	11.798		

Table 4.11 Multiple regression result for South Africa

South Africa	Multiple regression Result SDG Variable to growth rate		
	Coefficients	Standard Error	t Statistic
Intercept	−81.390	137.267	−0.592
Prevalence of Undernourishment SA	−0.660	0.825	−0.799
Life Expectancy at birth	1.926	0.851	2.263
MMR	0.007	0.017	0.458
NMR_SA	−2.677	0.929	−2.881
U5MR_IN	0.538	0.247	2.173
GER (%) Total upper secondary	−0.203	0.096	−2.107
GPI tertiary education	−27.621	16.253	−1.699
Decent jobs SA	0.3221	1.046	0.307
R^2	0.755		
F	4.249		

Discussion

Despite the negative relation between undernourishment and health indicators of South Africa, its performance in MMR, NMR and U5MR is better than India. To understand the phenomenon of why the data is not supporting the argument in case of health indicators in South Africa, I checked the government intervention in health sector. For doing this, I explored the expenditure on health in India and South Africa.

Health of an individual has always been considered as an index of human well-being. Access to healthcare services is recognized as fundamental right of an individual. The causal relation between health expenditure and poverty trap is an extensively discussed and well- researched topic across the regions (Nixon & Ulmann, 2006; Lulin Zhou et al., 2020; Taylor & Dessai, 2023; Biltagy & Hamdi, 2024; Thakur & Faizan, 2024). Causation between health and economic is well established (Doran & Kinchin, 2020; Ridhwan et al. 2022; Dhrifi & Jaziri, 2021).

Health expenditure is a key element of health system. The rate of mortality and morbidity are relatively higher among the poor population (Anser et al., 2020; Otavova et al., 2022). So, protection of poor from adverse health is part of welfare program. Health expenditure consist of public spending, out-of-pocket expenses, as well as health insurance provided under social welfare scheme and non-governmental organizations (WHO, 2019). World Health Organizations use domestic general government health expenditure (DGGHE) as percentage general government health expenditure (GGHE) to measure the public spending on health system. This indicator includes all government transfer and subsidies as well as contribution of social insurance scheme.

As shown in Table 4.12, contribution of DHCE in individual's health expenditure in India is very low compared to South Africa. In between 2015 and 2021, DGGHE in India and South Africa is almost stagnant, however the share of DGGHE in India was just 3.69, whereas the corresponding figure for South Africa was 15.29. The current health expenditure as percentage of GDP in South Africa is almost triple in comparison to South Africa (WHO, 2024).

When public expenditure on health system increases, it protects the people from poverty trap by seeking healthcare services. On contrary, poverty increase with high Out-of-pocket expense on health services. Out of pocket health expenditure refers to portion of health spending covered by people, through their own income, savings, and loan (World Bank, 2023a, 2023b, 2023c, 2023d, 2023e, 2023f, 2023g). People go through financial adversity in health when their out-of-pocket spendings impends their wellbeing or limits access to necessary commodities such as food, cloth, education or shelter (Getachew et al., 2023; Kolasa & Weychert, 2024). People who live on margin are being forced to choose between consumption of essentials goods emerged due to extra burden of health spendings. Evidences suggest that increase in health spending depletes savings or cause selling and borrowing (Kruk et al., 2009; Alshamsan et al., 2017). People are less likely to invest in future assets and their own human capital as well as their children (Dhanaraj, 2016).

The WHO data revealed that Indians out of pocket spending on health is nearly 10 times bigger than South African. As shown in Fig. 4.1, in 2021, an Indian shared nearly half of the total health expenditure over the same period a South African's bears just 5% of his total health expenditure. High dependence on Out-of-Pocket spendings for healthcare escalate financial stress of the households, many of them suffering catastrophic health spendings and impoverishment (2017). Subsequently, lack of health coverage force people to divert spendings to medical budget from essential commodities including education. It may be possible that due to high cost of health care, many Indians are less likely to consume health services, which resulted in low performance of India in health indicators as well as other social indicators.

World Bank statistics on Universal Health Coverage (UHC) shows that, in 2021, coverage of services under UHC in India was 63%, while during the same period coverage of UHC in South Africa was 71% (World Bank, 2023).

Table 4.12 Health Expenditures in India and South Africa

Health Expenditure	DGGHE % GGHE (%)		Current Health Expenditure (CHE) % of GDP (%)		Per Capita health spending US$ CHE	
	India	South Africa	India	South Africa	India	South Africa
2015	3.38	15.24	3.6	8.05	58	404
2016	3.39	15.29	3.5	8.08	–	–
2017	3.66	15.29	2.94	7.98	–	–
2018	3.68	15.29	2.86	8.09	–	–
2019	3.85	15.29	2.95	8.24	–	–
2020	3.69	15.29	3.34	8.57	–	–
2021	3.69	15.29	3.28	8.27	74	584

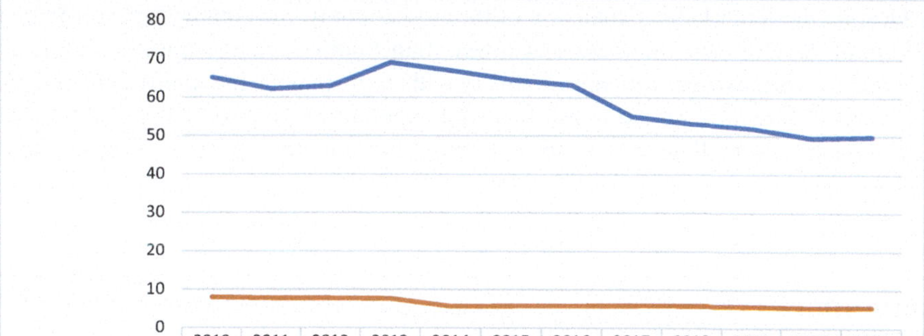

	2010	2011	2012	2013	2014	2015	2016	2017	2018	2019	2020	2021
India	65.18	62.22	63	69.07	67.01	64.66	63.21	55.11	53.23	52	49.45	49.82
South Africa	8	7.75	7.72	7.61	5.66	5.74	5.79	5.83	5.84	5.73	5.39	5.51

Fig. 4.1 Share of Out-of-pocket expenditure in current total health expenditure in India and South Africa. *Source* World Health Organization (2024)

In addition to high spending on health services, received cash assistance which South Africa have a comprehensive social grant system which covers almost half of the population (Rosenberg et al., 2024). The beneficiaries provide them freedom to channelize their spendings. It is reflected in per capita health expenditure which stood at US$ 504 in 2021. On the contrary, the poor in India are majorly supported by subsidized food grains. They have no freedom to choose the consumption basket which has led to low per capita spending on health covering merely US$ 74 over the same period (see Table 4.12).

Conclusion

This chapter examined the causality between GDP growth rate (proxy for well-being) and selected indicators of SDGs, which are envisaged in long term development targets in India and South Africa. The regression result demonstrates the high explanatory power of employed model, as selected indicators explain nearly 90% variation in GDP growth rate of India and almost 75% variation for GDP growth rate of South Africa.

The quantitative examination revealed that hunger and health indicators are closely associated with GDP growth rate in India with significant effect. On the other hand, in South Africa, hunger and health indicators are not substantive and need further investigation. Every indicator shows unique strength of correlation with GDP economic growth rate as well as within.

This result is significant for policy makers to identify the challenges of economy in the local preview and employ asymmetric approach to maximize the synergy, in contrast to the axiom that SDGs are indivisible. Indeed, economic growth and progress are important but other things are equally significant and deserve attention.

Despite the remarkable expansion of India's economy, the country's performance in human development indicators is lower than South Africa, whose economy is denoted by sluggish growth rate coupled with high rate of inequality and unemployment. While the positive trends in social expenditure give some hope, still there is a need for immediate intervention to shift the country on path of sustainable development.

Limitation

This study encountered few limitations related to availability of data for longer period. The study has employed nine variables. The predication power could be enhanced by adding more variables with longer time duration. Finally, the possibility of endogeneity related to time series data cannot be ruled out in present study.

References

Aiken, L. S., West, S. G., & Reno, R. R. (1991). *Multiple regression: Testing and interpreting interactions*. Sage Publication.

Alshamsan, R., Leslie, H., Majeed, A., & Kruk, M. (2017). Financial hardship on the path to Universal Health Coverage in the Gulf States. *Health Policy, 121*(3), 315–320. https://doi.org/10.1016/j.healthpol.2016.12.012

Ananad, S., & Sen, A. (1997). Concepts of human development and poverty: A multidimensional perspective. In A. Sen, & S. Ananad, *Poverty and Human Development: Human Development Papers 1997* (pp. 1–20.). United Nations Development Programme.

Anand, P., Jones, S., & Teitler, J. (2020). Non-monetary poverty and deprivation: A capability approach. *Journal of European Social Policy*, Available athttps://doi.org/10.1177/0958928720938334

Anand, S., & Ravallion, M. (1993). Human development in poor countries: on the role of private incomes and public services. *Journal of Economic Perspectives, 7*(1), es 133–150.

Anser, M. K., Yousaf, Z., Nassani, A. A., Alotaibi, S. M., Kabbani, A., & Zaman, K. (2020). Dynamic linkages between poverty, inequality, crime, and social expenditures in a panel of 16 countries: two-step GMM estimates. *Journal of Economic Structures, 9*, Article number 43, https://doi.org/10.1186/s40008-020-00220-6.

Badulescu, D., Simut, R., Badulescu, A., & Badulescu, A.-V. (2019). The relative effects of economic growth, environmental pollution and non-communicable diseases on health expenditures in european union countries. *International Journal of Environment Research Public Health, 16*(24), 5115, https://doi.org/10.3390/ijerph16245115. PMID: 31847367; PMCID: PMC6949912.

Benedek, D., Gemayel, E., Senhadji, A., & Tieman, A. F. (2021). *Achieving the sustainable development goals will require extraordinary effort by all.* Available on: https://www.imf.org/en/Blogs/Articles/2021/04/29/blog-achieving-the-sustainable-development-goals: IMF.

Biltagy, M., & Hamdi, M. (2024). Public health expenditure and household poverty: Case study of Egypt. *Future Business Journal, 10*, Article number 7, https://doi.org/10.1186/s43093-023-00295-0.

Blanc, D. L. (2015). Towards Integration at Last? The Sustainable Development Goals as a Network of Targets. *Sustainable Development*, 176–187. Available at: https://doi.org/10.1002/sd.1582.

Bourguignon, F., & Chakravarty, S. (2003). The measurement of multidimensional poverty. *The Journal of Economic Inequality*, 25–48. Avaialble at: https://doi.org/10.1023/A:1023913831342.

Cerf, M. E. (2019). Sustainable development goal integration, interdependence, and implementation: The environment–economic–health nexus and universal health coverage. *Global Challenges*, Wiley online. Available at: https://doi.org/10.1002/gch2.201900021.

Cerf, M. E. (2023). The social-education-economy-health nexus, development and sustainability: Perspectives from low- and middle-income and African countries. *Discover Sustainablity, 4*, Article number 37, https://doi.org/10.1007/s43621-023-00153-7.

Cerra, M. V., Lama, M. R., & Loayza, N. (2021). Links between growth, inequality, and poverty: A survey. *IMF Working Paper* (68), https://doi.org/10.5089/9781513572666.001.

Chicco, D., Warrens, M. J., & Jurman, G. (2021). The coefficient of determination R-squared is more informative than SMAPE, MAE, MAPE, MSE and RMSE in regression analysis evaluation. *PeerJ Computer Science, 7.* https://doi.org/10.7717/peerj-cs.623 PMID: 34307865

Cling, J.-P., & Delecourt, C. (2022). Interlinkages between the sustainable development goals. *World Development Perspectives, 25*, Available at: https://doi.org/10.1016/j.wdp.2022.100398.

D.Sachs, J. (2001). *Macroeconomics and health: Investing in health for economic development.* https://iris.who.int/bitstream/handle/10665/42435/924154550X.pdf?amp;isAllowed=y&sequence=1: World Health Organization.

Dhanaraj, S. (2016). Effects of parental health shocks on children's schooling: Evidence from Andhra Pradesh. *India. International Journal of Educational Development., 49*, 115–125. https://doi.org/10.1016/j.ijedudev.2016.03.003

Dhrifi, A., & Jaziri, S. A. (2021). The causal links among economic growth, education and health: evidence from developed and developing countries. *Journal of the Knowledge Economy, 12*, 1477–1493. https://doi.org/10.1007/s13132-020-00678-6.

Dissanayake, S., & Mahadev, R. (2020). Evaluating the efficiency of carbon emissions policies in a large emitting developing country. *Energy Policy, 136*, Available on: https://doi.org/10.1016/j.enpol.2019.111080.

Doran, C. M., & Kinchin, I. (2020). Economics of mental health: Providing a platform for efficient mental health policy. *Applied Health Economics and Health Policy, 18*, 143–145. https://doi.org/10.1007/s40258-020-00569-6

Fajnzylber, P. (2018). *Why growth alone is not enough to reduce poverty.* Available at: https://ieg.worldbankgroup.org/blog/why-growth-alone-not-enough-reduce-poverty: World Bank.

FAO, IFAD, UNICEF, WFP & WHO. (2023). *The state of food security and nutrition in the world 2023. Urbanization, agrifood systems, transformation and healthy diets across the rural-urban contnuum.* FAO.

FAO, IFAD, UNICEF, WFP, WHO. (2022). Food security and nutrition around the world. In *The State of Food Security and Nutrition in the World 2022* (pp. 9–47). FAO. Available at: https://doi.org/10.4060/cc0639en.

Farrar, D. E., & Glauber, R. R. (1967). Multicollinearity in regression analysis: The problem revisited. *The Review of Economics and Statistics, 49*(1), 92–107.

Future and potential spending on health 2015–40: development assistance for health, and government, prepaid private, and out-of-pocket health spending in 184 countries. (2017). *389*(10083), 2005–2030.

Getachew, N., Shigut, H., Edessa, G. J., & Yesuf, E. A. (2023). Catastrophic health expenditure and associated factors among households of non community based health insurance districts, Ilubabor zone, Oromia regional state, southwest Ethiopia. *International Journal of Equity Health, 22*, Article number 40, https://doi.org/10.1186/s12939-023-01847-0.

Grossman, M. (1972). On the concept of health capital and the demand for health. *Journal of Political Economy, 80*(2), 223–255. http://www.jstor.org/stable/1830580.

Ho, S. S., Bowser, G., Templer, P., & Green, S. A. (2023). Learning for sustainability: Partnerships for the goals. *Sustainable Earth Reviews, 6*, 8. https://doi.org/10.1186/s42055-023-00059-2

Huang, Y., Huang, X., Li, R., & Cheng, W. (2023). Assessing the effectiveness of targeted poverty alleviation policies in Xinjiang, China. *Humanities and Social Sciences Communications, 10*, Article number 389, Available at: https://doi.org/10.1057/s41599-023-01858-w.

IMF. (2024). *IMf DATA Mapper: Data 2024.* Available at: https://www.imf.org/external/datamapper/NGDPD@WEO/IND/GBR: International Monetary Fund.

ISC. (2017). *A guide to sdg interactions: From science to implementation.* https://council.science/publications/a-guide-to-sdg-interactions-from-science-to-implementation/: International Science Council.

Janus, H., & Keijzer, N. (2013). *Post-2015: How to Design Goals for (Inter)National Action?* German Development Institute Briefing Paper No. 23/2013.

Knijnenburg, T. A., Wessels, L. F., Reinders, M. J., & Shmulevich, I. (2009). Fewer permutations, more accurate P-values. *Bioinformatics, 25*(12), 161-i168.

Kolasa, A., & Weychert, E. (2024). The causal effect of catastrophic health expenditure on poverty in Poland. *The European Journal of Health Economics., 25*, 193–206. https://doi.org/10.1007/s10198-023-01579-6

Kruk, M. E., Goldmann, E., & Galea, S. (2009). Borrowing and selling to pay for health care in low-and middle-income countries. *Health Aff (millwood), 4*, 1056–1066. https://doi.org/10.1377/hlthaff.28.4.1056

Livieris, I. E., Stavroyiannis, S., Iliadis, L., & Pintelas, P. (2021). Smoothing and stationarity enforcement framework for deep learning time-series forecasting. *Neural Computer Appllication, 33*(20), 14021–14035, https://doi.org/10.1007/s00521-021-06043-1.

Livieris, I. E., Stavroyiannis, S., Pintelas, E., & Pintelas, P. (2020). A novel validation framework to enhance deep learning models in time-series forecasting. *Neural Computing and Applications., 32*, 17149–17167. https://doi.org/10.1007/s00521-020-05169-y

Lukman, A. F., Farghali, R. A., Kibria, B. M., & Oluyemi, O. A. (2023). Robust-stein estimator for overcoming outliers and multicollinearity. *Scientific Reports, 13*, Article number 9066, https://doi.org/10.1038/s41598-023-36053-z.

Lulin Zhou, S. A.-W., Xu, X., Salman, M., Antwi, M. O., & Afua, T. M. (2020). An empirical study on the determinants of health care expenses in emerging economies. *BMC Health Services Research, 20*, Article number 774, https://doi.org/10.1186/s12913-020-05414-z.

Lusseau, D., & Mancini, F. (2019). *Nature Sustainability, 2*, 242–247.

Luttikhuis, N., & Wiebe, K. S. (2023). Analyzing SDG interlinkages: Identifying trade-offs and synergies for a responsible innovation. *Sustainabilty Science, 18*, 1813–1831. https://doi.org/10.1007/s11625-023-01336-x

Mackie, E. (2012). Understanding Poverty: The human development approach. *E-International Relation*, pp. https://www.e-ir.info/2012/04/17/understanding-poverty-the-human-development-approach/.

Mason, G. (1987). Coping with collinearity. *Canadian Journal of Program Evaluation > List of Issues >, 2*(1), 87–93.

McGowan, P. J., Stewart, G. B., Long, G., & Grainger, M. J. (2019). An imperfect vision of indivisibility in the Sustainable Development Goals. *Nature Sustainabilty*, 43-45.https://doi.org/10.1038/s41893-018-0190-1

McKinsey & Company. (2023). Unleashing India's full potential. *McKinsey & Company*, pp. Available at: https://www.mckinsey.com/featured-insights/themes/unleashing-indias-full-potential.

Ministry of Education. (2023). *Education for all in India.* Available at: https://educationforallinindia.com/role-of-gender-parity-index-in-higher-education-in-india-2023/: Ministry of Education, Government of India.

Nature. (2023). EDITORIAL: GDP at 70: why genuinely sustainable development means settling a debate at the heart of economics. *Nature*, https://www.nature.com/articles/d41586-023-02509-5.

Nilsson, M., Griggs, D., & Visbeck, M. (2016). Policy: Map the interactions between Sustainable Development Goals. *Nature*, 320–322. https://www.nature.com/articles/534320a.

Nixon, J., & Ulmann, P. (2006). The relationship between health care expenditure and health outcomes: Evidence and caveats for a causal link. *The European Journal of Health Economics, 7*(1), 7–18.

NPC SA. (2011). *National Development Plan-2030.* National Planning Commission. Republic of South Africa. Available at: https://www.nationalplanningcommission.org.za/National_Development_Plan.

OECD & UNDP. (2019). *G20 contribution to the 2030 Agenda: PROGRESS AND WAY FORWARD.* www.oecd.org/dev/OECD-UNDP-G20-SDG-Contribution-Report.pdf: Organisation for Economic Co-operation and Development and the United Nations Development Programme.

Omar, M. A., & Inaba, K. (2020). Does financial inclusion reduce poverty and income inequality in developing countries? A panel data analysis. *Journal of Economic Structures, 9*, Article number 37, Available at: https://doi.org/10.1186/s40008-020-00214-4.

Otavova, M., Faes, C., Bouland, C., Clercq, E. D., Vandeninden, B., Thierry Eggerickx, J.-P. S., ... Masquelier, B. (2022). Inequalities in mortality associated with housing conditions in Belgium between 1991 and 2020. *BMC Public Health, 22*, Article number 2397, https://doi.org/10.1186/s12889-022-14819-w.

Park, C.-Y., & Mercado, R. (2018). Financial inclusion, poverty, and income inequality. *The Singapore Economic Review, 63*(1), 185–206. Available at: https://doi.org/10.1142/S0217590818410059.

Payab, A. H., Kautish, P., Sharma, R., Siddiqui, A., Mehta, A., & Siddiqui, M. (2023). Does human capital complement sustainable development goals? Evidence from leading carbon emitter countries. *Utilities Policy, 81*, Available at: https://doi.org/10.1016/j.jup.2023.101509.

PIB. (2023). Budget 2023–24 Presents vision for Amrit Kaal- Blue Print for an Empowered and Inclusive Economy. *Press Bureaue of India*, p. https://pib.gov.in/PressReleaseIframePage.aspx?PRID=1895313.

Pradhan, P., Costa, L., Rybski, D., & W Lucht, J. K. (2017a). A systematic study of sustainable development goal (SDG) interactions. *Advancing Earth and Space Science*, Available onhttps://doi.org/10.1002/2017EF000632

Pradhan, P., Costa, L., Rybski, D., Lucht, W., & Kropp, J. P. (2017b). A Systematic study of sustainable development goal (SDG) interactions. *Earth's Future*, Available athttps://doi.org/10.1002/2017EF000632

Pradhan, P., Costa, L., Rybski, D., Lucht, W., & Kropp, J. P. (2017c). A systematic study of sustainable development goal (SDG) interactions. *Earth's Future*, Availble athttps://doi.org/10.1002/2017EF000632

Rao, N. D., & Min, J. (2018). Less global inequality can improve climate outcomes. *Wiley Interdisciplinary Reviews: Climate Change, 9*(2), Avaialble at: https://doi.org/10.1002/wcc.513.

Ridhwan, M. M., Nijkamp, P., Ismail, A., & M.Irsyad, L. (2022). Masagus M. Ridhwan, Peter Nijkamp, Affandi Ismail & Luthfi M.Irsyad. *Empirical Economics, 63*, 3211–3251. https://doi.org/10.1007/s00181-022-02226-4.

Rodrik, D. (2000). Growth versus poverty reduction: A hollow debate. *Finance and development, 37*(4), Available at: https://www.imf.org/external/pubs/ft/fandd/2000/12/rodrik.htm.

Rolfe, S., Garnham, L., Godwin, J., Anderson, I., Seaman, P., & Donaldson, C. (2020). Housing as a social determinant of health and wellbeing: developing an empirically-informed realist theoretical framework. *BMC Public Health, 20*, Article number 1138, Available at: https://doi.org/10.1186/s12889-020-09224-0.

Rosenberg, M., Kabudula, C. W., Kahn, K., & Kobayashi, L. (2024). Money and ageing: South African study shows cash grants help people live longer and have better memory function. pp. https://theconversation.com/money-and-ageing-south-african-study-shows-cash-grants-help-people-live-longer-and-have-better-memory-function-222726.

Salecker, L., Ahmadov, A. K., & Karimli, L. (2020). Contrasting monetary and multidimensional poverty measures in a low-income sub-Saharan African Country. *Social Indicators Research., 151*, 547–574.

Scherer, L., Behrens, P., Koning, A. d., Heijungs, R., Sprecher, B., & Tukker, A. (2018). Trade-offs between social and environmental sustainable development goals. *Environmental Science & Policy, 90*, 65–72.https://doi.org/10.1016/j.envsci.2018.10.002

Schultz, T. W. (1972). *Human capital: Policy issues and research opportunities*. NBER. Available at:: http://www.nber.org/books/schu72-1.

Sen, A. (2000). *Development as freedom*. Anchor.

Sen, A. (2002). Why health equity? *Health Economics*, Wiley Online. https://doi.org/10.1002/hec.762.

Sen, A. (2011). *Idea of Justice*. Belknap Press: An Imprint of Harvard University Press.

Siddiqui, F., A, R., Salam, Lassi, Z. S., & Das, J. K. (2020). The intertwined relationship between malnutrition and poverty. *Front Public Health, 8*, 453, Available at: https://doi.org/10.3389/fpubh.2020.00453.

Singh, U. K. (2022). *Human development in Bihar: Experiences of Millennium developement goals*. Springer.

Singh, U. K. (2023). Realization of Amrit Kaal through Sustainable Development Goals. *Financial Express*, pp. https://www.financialexpress.com/business/defence-realization-of-amrit-kaal-through-sustainable-development-goals-2996437/.

Slim, H. (1999). What Is Development? *What Is Development? Development in Practice, 5*(2), 143–148. Available at: https://www.jstor.org/stable/4028933.

Smith, T. B., Vacca, R., Mantegazza, L., & Capua, I. (2021). Natural language processing and network analysis provide novel insights on policy and scientific discourse around Sustainable Development Goals. *Scientific Reports, 11*, Article number 22427, Available at: https://doi.org/10.1038/s41598-021-01801-6.

Smith, T. B., Vacca, R., Mantegazza, L., & Capua, I. (2023). Discovering new pathways toward integration between health and sustainable development goals with natural language processing and network science. *Globalization and Health, 19*, Article number 44, Available at: https://doi.org/10.1186/s12992-023-00943-8.

STATSA. (2015). *Millennium Development Goals: Country report 2015*. Available at: https://www.statssa.gov.za/MDG/MDG_Country%20Report_Final30Sep2015.pdf: Statistics South Africa.

Taylor, F., & Dessai, S. P. (2023). Relation between per capita public expenditure on health and income in India at sub-national level. *Indian Journal of Public Administration, 69*(2), 317–329. https://doi.org/10.1177/00195561221149914

Thakur, R., & Faizan, M. A. (2024). Magnitude of health expenditure induced removable poverty in India: Some reflections of Ayushman Bharat. *Heliyon., 10*(1), E23464. https://doi.org/10.1016/j.heliyon.2023.e23464

Tosun, J., & Leininger, J. (2017). What are SDG Interlinkages and why are they relevant? *Global Challenges*, Wiley Online Library.https://doi.org/10.1002/gch2.201700036

Ugnich, E., Chernokozov, A., & Ugnic, M. (2021). Human capital in the system of sustainable development goals: significance and prospects. *E3S Web of Conferences 258, 07053*, 1–7. Available at: https://doi.org/10.1051/e3sconf/202125807053.

UIS STAT. (2023). *Gross enrolment ratio by level of education*. Available at: http://data.uis.unesco.org/index.aspx?queryid=3812: UNESCO.

UN. (1990). *Human development report 1990*. United Nations Development Program.

UN. (1995). Eradication of Poverty. In UN, *Programme of Action of the World Summit for Social Development* (pp. Available at: https://www.un.org/development/desa/dspd/world-summit-for-social-development-1995/wssd-1995-agreements/pawssd-chapter-2.html). United Nations.

UN. (2015). *Resolution adopted by the General Assembly on 25 September 2015*. United Nations.

UN. (2023). *Take action for the sustainable development goals*. Available at: https://www.un.org/sustainabledevelopment/sustainable-development-goals/: United Nations.

UN. (2024). *Nutituion and Food Security*. Avaialble at: https://india.un.org/en/171969-nutrition-and-food-security: UNited Nations Digital Library.

UNDP. (2023a). *Global multidimensional poverty index (MPI) 2023*. United Nations Debvelopment Programme.

UNDP. (2023b). *Human development report 2023.* United Nations Development Programme.

UNESCO. (2023). *Global education monitoring report 2023: Technology in education-A tool on whose terms?* UNESCO. https://doi.org/10.54676/UZQV8501:Paris

UNFPA. (2023). *World population dashboard.* Avaialble at: https://www.unfpa.org/data/world-population-dashboard.

UNICEF. (2023). *Maternal mortality.* UNICEF. Available at: https://data.unicef.org/topic/maternal-health/maternal-mortality/.

Vatcheva, K. P., Lee, M., Cormick, J. B., & Rahbar, M. H. (2016). Multicollinearity in regression analyses conducted in epidemiologic studies. *Epidemiology (Sunnyvale) 6*(2), 227, Available at: https://doi.org/10.4172/2161-1165.1000227.

Wheebox. (2023). *India Skills Report 2023: Roadmap to India's Skill and Talent Economy 2030.* Avaialable at: https://do3n1uzkew47z.cloudfront.net/siteassets/pdf/ISR_Report_2023.pdf: Wheebox.

WHO. (2019). *people are still paying too much out of their own pockets.* World Health Organoizations.

WHO. (2024). *Out-of-pocket expenditure as percentage of current health expenditure (CHE) (%).* https://www.who.int/data/gho/indicator-metadata-registry/imr-details/4965: World Health Organization.

Worl Bank. (2019). *Implementing the Agenda 2030: 2018 Update.* Avaialble on: https://pubdocs.worldbank.org/en/893311531748303554/2030Agenda-2018Update-July-19.pdf: World Bank.

World Bank. (2023a). *Mortality rate, under-5 (per 1,000 live births)—India, South Africa.* Available at: https://data.worldbank.org/indicator/SH.DYN.MORT?end=2021&locations=IN-ZA&start=2000: World Bank.

World Bank. (2023b). *Poverty and Inequality Platform: Country Profile India.* Available at: https://pip.worldbank.org/country-profiles/IND: World Bank .

World Bank. (2023c). *Tracking universal health coverage: 2023 global monitoring report.* World Bank.

World Bank. (1989). *Poverty & Human Development.* World Bank. Available at: https://documents1.worldbank.org/curated/en/838741468766811654/pdf/multi0page.pdf.

World Bank. (2022). *Global poverty monitoring technical note: September 2022 update to the poverty and inequality platform (PIP) What's New.* Available at: IDU05b43a261041c504a5f0bb3405d0ef310b9e1.pdf: The World Bank.

World Bank. (2023d). *Mortality rate, neonatal (per 1,000 live births)—India, South Africa.* Available at: https://data.worldbank.org/indicator/SH.DYN.NMRT?end=2021&locations=IN-ZA&start=2000: World Bank.

World Bank. (2023e). *Prevalence of undernourishment (% of population)—India.* Available on: https://data.worldbank.org/indicator/SN.ITK.DEFC.ZS?locations=IN: World Bank.

World Bank. (2023f). *School enrollment, tertiary (gross), gender parity index (GPI)—India, South Africa.* Available at: https://data.worldbank.org/indicator/SE.ENR.TERT.FM.ZS?end=2021&locations=IN-ZA&start=2000: World Bank.

World Bank. (2023g). *Tracking Universal health coverage: 2023 global monitoring report.* World Bank.

Wright, S. (1921). Correlation and causation. *Journal of Agricultural Research, 20*(3), 557–585.

Zhang, S., Yu, Y., Kharrazi, A., Ren, H., & Ma, T. (2022). Quantifying the synergy and trade-offs among economy–energy–environment–social targets: A perspective of industrial restructuring. *Journal of Environmental Management, 316*,https://doi.org/10.1016/j.jenvman.2022.115285

Zhao, W., Yin, C., Hua, T., Meadows, M. E., Li, Y., Liu, Y., . . . Fu, B. (2022). Achieving the sustainable development Goals in the post-pandemic era. *Humanities and Social Sciences Communications, 9*, Article number 258, Available at: https://doi.org/10.1057/s41599-022-01283-5.

Zhu, Y., Bashir, S., & Marie, M. (2022). Assessing the relationship between poverty and economic growth: Does sustainable development goal can be achieved? *Environmental Science and Pollution Research, 29*, 27613–27623. Avaialble at: https://doi.org/10.1007/s11356-021-18240-5.

Chapter 5
Conclusion and Way Forward

Introduction

The idea of human development is encapsulated in the promise of freedom made by the founding fathers of India and South Africa; freedom as a step forward to eradicate poverty and ignorance, reducing the burden of diseases, and ensuring equal and unbiased opportunities for all. To safeguard the promises, the constitution of India and South Africa ensures political freedom as well as access to those goods essential for human well-being.

This cross-country research has compared the political dimensions and traversed the stages in public policy from planning to implementation and from reach to range in the quest for human development in India and South Africa. There are many reasons for the comparative study of emergent economies such as India, South Africa, China, Brazil, and Russia. The share of emerging economies in total world's GDP increased from 50% in 2005 to 60% in 2023 (IMF, 2023a, 2023b). This implies that advanced economies such as the United States, the European Union, and Japan are no longer masters of the world economy (Widrick, 2007). Dadush and Shaw Shaw (2011) projected this development as the transformation of the developing world from supplicants to trendsetters. Srivastav (2012) noted that the emergence of China, India, and South Africa has changed the picture of political economy. Regional blocks like BRICS are giving intense competition to Western hegemony (Becker, 2014). According to a global consulting firm, the power shift from advanced economies will continue till 2050 (PWC, 2017). The emergence of developing economies indicates that their role in the global economy is more than that of a supplier of labor-intensive goods, but that they are a global growth engine (Joshi & Klein, 2018). Their share continues to boost the world's GDP, despite uneven performance in recent years. The rise in the economic capacity of emerging market economies empowers them to exert more bargaining weight in the global economic order and decision-making process (Nuruzzaman, 2020). Wu and Pan (2021) observed the significant contribution of emerging economies to global economic growth, especially Asian countries during

U. K. Singh, *Regional Dimensions of Human Development in India and South Africa*,
Advances in Geographical and Environmental Sciences,
https://doi.org/10.1007/978-981-97-5846-3_5

the few last decades. Liu (2021) concluded that the rise of emerging economies has a great implication in shaping global governance. On the contrary, Dani Rodrik (2024) argued that in the neoliberal era world has witnessed the expansion of the global economy, while the widening inequality within countries, deteriorating climate, and issues of public health attract attention.

Although India has achieved stupendous economic growth over the recent decades, the country performed weakly in the manufacturing sector. Despite the improvement in poverty reduction, India is still home to the largest poor population in the world. The daunting poverty indicates that drivers of upliftment such as access to good health, quality education, skill, and access to clean and sustainable water for drinking purposes and sanitation services are not reaching everyone. On the contrary, despite the sluggish economic growth rate, South Africa performed better than India in Human Development Index ranking. The Global Hunger Index shows the miserable condition of India as India stood at 111 ranks out of 125 countries, while South Africa was placed 58 (Grebmer et al., 2023). Notwithstanding the highly industrialized economy, South Africa faces the likewise challenge of high unemployment, poor quality of primary education, low skilled labour force, poverty, and persistent inequality.

Drawing an inference from the quantitative analysis of Chap. 4, I conclude that the idea of human well-being is linked to capability enhancement by accumulating stocks of health and education. Income provides access to health and education and other capabilities, so expansion of capabilities is anticipated with the rise in income. Several recent studies in the recent decade indicate that investment in education improves when returns in education are higher (Leoni, 2023; Wolla & Vandenbroucke, 2023; Psacharopoulos & Patrinos, 2023; Shah & Steinberg, 2017; Bharadwaj et al., 2019). The idea is widely discussed in Chap. 1.

Assessment of First Half of Sustainable Development Goals in India and South Africa

This section provides the assessment of SDGs in the first half of its implementation as inferred from Chaps. 2 and 3.

SDG1: No Poverty

Descending multidimensional poverty level in India was applauded by UNDP in its Multidimensional Poverty Report 2023 which indicated that almost 415 million people escaped poverty between 2005/06 and 2019/2. The proportion declined to 164% in 2023. However, the report was based on data collected in pre-pandemic survey and the pandemic has exacerbated the global count of multidimensional poverty (UN, 2023). In India, nearly 230 million people were pushed into poverty due to the pandemic (Basole et al., 2021). It is nearly impossible to identify the share of the population living in extreme poverty because the country has abandoned poverty estimation and relies on Multidimensional data for policy measures. The proportion of multidimensional poor in South Africa is just 6.3% over the same period. South Africa has a unique approach to poverty estimation at the national level based on income level as well as access to food. The poverty level ranges between nearly 25% (food poverty), 40% (lower-bound poverty), and 56% (upper-bound poverty in 2020 (STATS SA, 2023). South Africa with its ample data sources has set up overarching strategies to protect people from vulnerability. World Bank's (2022) readings show that extreme poverty during COVID-19 has been contracted in South Africa.

SDG2: Zero Hunger

Despite the self-sufficiency in food grains at the national level, hunger is a serious policy issue in India and South Africa (UN, 2022; STATS SA, 2021). A large proportion of people in India and South Africa are undernourished. India is home to nearly 21.5 million (18.7%) children affected by wasting, the highest in the world, and 36.1 million (31.7%) children who are stunting, whereas in South Africa the counts of undernourished people rose from 1.8 million in 2006/08 to 4.7 million malnourished people in 2020/22, with a mere decrease between period 2013/15 and 2016/18. Every third child (under the age of five) is stunted, and the number goes to 1.5 million. Every second child death is linked to chronic malnutrition among children. Nearly 75,000 children are suffering from obesity.

Hunger population in both countries are covered under one of the world largest social security systems. In India, nearly 60% of the population, accounting for 813 million people survive on subsidized food grain provided under the central government scheme. South Africa chose to address hunger through the social security system, the largest among its peers. To some extent, SSA is also the main source of income for many South Africans.

SDG3: Health and Wellbeing

India and South Africa have shown mixed progress in health indicators. Overburdened public hospitals and primary healthcare units are facing a shortage of trained health professionals. India has achieved its national SDG target achieved its national target of reducing the MMR to 100/100000 live births, by reducing the number to 97/ 100,000 live births (PIB, 2023), whereas recent data for South Africa was 127 per 100,000 live birth (UNICEF, 2023) and is likely to achieve the target of 70 deaths per 100,000 live births by 2030.

The indicators of Neo-Natal Mortality Rate (NMR), Under-five Mortality Rate (U5MR) shows considerable improvement in India and South Africa. In India, the number of NMR and U5MR declined from 57/1000 births to 19/1000 births and 127/1000 births to 31/1000 births respectively during 1990 and 2021. While South Africa South has achieved the target of NMR far back in 2011, the readings for U5MR in 2020 was 33.7 deaths per 1000, the country is likely to achieve the target of 25 deaths per 1000 live births. The cases of HIV reduced to 1/1000 in age group 1 per 1000 people. However, recent trends have shown an increase in alcohol consumption which negatively affects other health targets.

SDG4: Quality Education

India and South Africa have shown impressive progress in keeping pupils in school along with successful completion of their grades. South Africa has achieved high rate of secondary school pass-out with 94.2% in 2021 from 73% in 2002, the corresponding figure for senior secondary was 61.3% during the same period. However, in India, 98.1% students in primary (1 to 5), 96.3 in upper primary (6 to 8), and 86.3 (9–10) have passed their grades. In both countries, the trends decrease with an increase in grades.

The improvement in pre-university education also translates into progress in access to tertiary education. However, enrolment in tertiary education is very low in comparison to primary and secondary education. Almost half of Indian youth (51.25%), lack skills for prevailing employment (Wheebox, 2024). Employability among South African youth is much lower with nearly one-third of youth (15–24 years) falling under the category 'not in education, employment and Training (NEET).

There may be several reasons for the low participation of Indian and South African youths in higher education, for certain, poor quality of primary education, and financial constraints. Providing skill training is a major challenge in both countries.

SDG5: Gender Equality

South Africa and India shows opposite figures of women's proportion in decision making. South Africa has nearly achieved the target of fifty percent women in national parliament while the same progress is not achieved in second and third tier of democracy. While India has achieved the parity in decision making in third tier, but still lag behind in second and first tier.

Women's labor force participation in South Africa is better than India. Recent trends for India show that the women are leaving the labor force.

SDG6: Clean Water and Sanitation

In India nearly 35% dwellings are having supply of piped water while 95% are using improved source of water, whereas almost 83% South African households are receiving piped water. However, the quality of water is debatable. The water may be clean, free form physical pollutant but contaminated with heavy metals and chemical, unhealthy for drinking purpose.

SDG7: Affordable and Clean Energy

Despite government subsidy one third of Indian household rely on primary sources of energy (firewood, chips and crops residue) for cooking fuel. Nevertheless, rural to urban area divide is stark, 46.5% in rural area over 6.5 in urban cohort, and 33.8% at the national level. However, 89% in urban areas and 49.4% in rural areas used LPG for cooking purposes. While in South Africa 77% of households use modern energy for cooking purposes (MECS, 2023). However, in India, people have gas cylinders (clean fuel for cooking) but they are still dependent on fossil fuels for cooking and other house chores because of financial constraints. India and South Africa are investing in green energy. India has made significant progress in green energy investment; South Africa has also planned to start new nuclear power plants.

SDG8: Decent Jobs

Decent job is a major challenge in both countries. Both countries are facing unprecedented youth unemployment. However, job markets in India and South Africa differ in nature based on their economic structure, and common skill mismatch between existing jobs and labor force.

SDG10: Rising Inequality

South Africa is the most unequal country in the world, and the rising inequality in India also placed it in the club's most unequal country. The stark inequality between the top 1% and the bottom 50% is a major concern for policymakers in India and South Africa.

Policy Assessment in Context of Human Development in India and South Africa

Social Security System

Despite progress in human development indicators, poverty, inequality and unemployment especially among the youth remains the major concern for policy makers. IMF study on Caucasus region indicated that social security system (SSS) can be effective in reducing poverty, inequality and unemployment (Gigineishvili et al., 2023). SSA is rapidly taking space in policy design of developing countries. These programs are non-contributory, designed to protect the people from adversity and vulnerability (2012). According to World Bank (2019), nearly 2.5 billion people across the world covered by SSS programs and roughly 650 million (56%) of the poorest quintiles.

The ongoing social security programs of poverty alleviation includes, unconditional cash transfers, pensions, food subsidies, agriculture subsidies, direct credit support, asset creations, and public works. In a very recent MIT research paper, Banerjee et al. (2023) outlined that at present more than 120 countries have implemented cash transfer programs, out of which nearly 70 countries run social pension schemes. World Bank (2018a, 2018b) noted that SSA has the potential to simultaneously enhance equity and promote equal opportunity by uplifting people and society from destitution. The targeted SSS programs mitigate the negative economic shocks and natural calamities on people and society. Overarching policies create opportunities by strengthening human capital and enabling access to employment and investment in people's day-to-day lives and livelihoods. A real-time study of the World Bank shows that the COVID-19 pandemic has given a further push to SSS programs, indicating their increased role in future policy (Gentilini et al., 2022). Barrientos (2013) argued that social protection programme focusses on the vulnerable and marginalized population through direct benefit scheme.

The SSS programs are strategized for redistributive purposes. The successful implementation of SSS depends on state's capacity, institutional efficiency and resources. Hanna and Karlan (2017) recognizes that the poor's marginal utility of consumption is higher than for the rich, and therefore make it socially optimal through a perspective of utilitarians (on axiom that the taxation process does not generate significant dead weight losses). While the structural design of these programs differs from one country to another, they share the common objective of identifying the persons in need along a certain criterion, further benefiting them with SSS programs. If the amount of cash transfer is large enough to offer appropriate capital, either to start a new venture, implement new agriculture practices (Banerjee & Newman, 1993), they could accelerate economic growth in the long run, or encourage risk averse people to invest in precarious with profitable activities (Chetty & Looney, 2006) and move forward. But, the main objective of SSS programs is to reduce poverty and hunger and ensure minimum living standard of an individual. However, the major challenge of these initiative is policy target, reach to the neediest person.

Despite several strategic attempts to reduce poverty by local, national and global policy makers, poverty continues to exist and pose a serious threat to humanity. Existing literature finds close correlation between poverty, poor health, and unemployment (Nilsson, 2015; Etana & Tolossa, 2017; Pohlan, 2019; AbhijitBanerjee, EstherDuflo, & GarimaSharma, 2021). Gallie, Paugam and Jacobs' (2010) study shows that unemployment has adverse effect on poverty as well as health. The chances to slip in poverty increases with unemployment and inactivity. On the contrary, employment provides an opportunity to exit poverty (Vaalavuo & Sirniö, 2021).

Rising unemployment among Indian youths and persistently high rates of unemployment across all age groups raise a grave policy concern. Employment is the way to exit poverty for people in India and South Africa. However, some working populations in India and South Africa are still considered poor depending on the nature of work. Employment can contain the probability of being poor. Job is a means to empowerment. Income through jobs improves access to basic necessities essential for well-being. On the other hand, unemployment raises frustration and has a negative influence on men, pushing them into illegal activities.

The above discussion points out that creating employment through public works has the potential to raise the income of people. This salient feature of public works schemes has been taking place in policy design worldwide. In a cross-country research, Tillin and Duckett (2017) outlined that emergent economies such as Brazil, China, India, and South Africa have designed a new set of welfare programs to address poverty and inequality in contrast to global competitiveness in the globalized world of the twenty-first century.

Expanded Public Work Programme (EPWP) in South Africa and Mahatma Gandhi National Rural Job Guarantee Act-2005 in India are examples of low-wage employment. An intervention to mainstreaming poor, marginalized, and vulnerable groups, and enables unskilled and unemployed people to propel the engine of sustainable development. MGNREGA and EPWP are extensively discussed in Chaps. 2 and 3.

MGNREGA and EPWP: The Employer of Last Man in the Queue

MGNREGA is a legally entitled program that engaged 28% of rural Indian households during 2019–2020, nearly 76 million people benefited over the same period (Narayan, 2022). During and after the COVID-19 pandemic, MGNREGA proved as employer of the last man in the queue. Human Development Report (UNDP, 2015a, 2015b) finds MGNREGA as one of the most effective social security interventions to tackle poverty. In the initial phase, MGNREGA lowered poverty by a third. It has become the largest scheme in Independent India to provide income in cash, especially to rural Indian women (2021). It played a significant role in preventing people from falling into poverty. In contrast to other public work schemes in developing countries, the MGNREGA has been extensively studied by economists and other academic researchers. Among them, Imbert and Paap's (2019) study established that income through MGNREGA helped in limiting rural to urban migration. Based on a vast survey of 209 Indian districts, across 18 states, Berg et al. (2018) study suggested that the act has increased daily wages in agricultural jobs by 4.3% per annum. The benefits of MGNREGA are more profound in downtrodden communities such as Schedule Caste, and women-headed households (Deininger & Liu, 2019). World Bank literature argued that MGNREGA works are free from any environmental shocks, and has the potential to contain income uncertainty due to crop failure (Gehrke, 2019). Fetzer (2020) in his research held that MGREGA has effectively weakened the link between income and productivity shock due to externalities. A very recent ILO literature noted that this act has significantly contributed to the rise of wages for rural workers (Ananian & Dellaferrera, 2024). However, these benefits are temporary and not sustainable in the long term.

In the absence of regular jobs, unemployed Indians are looking towards MGNREGA for immediate relief. Various research highlighted that the short-term relief from MGNAREGA have a negative impact on human capital and the rise in low-wage jobs has a negative impact on investment in human capital. A Harvard study shows that wage increment in the labor market decreases school attendance among older children (Islam & Sivasankaran, 2015). Li and Sekhri (2020) finding suggest that enrollment in schools decreases and quality and quality of education deteriorate due to MGNREGA, this is further complemented by the study of Shah and Steinberg (2017), who showed that the investment in human capital decreases with increase in child's age. He found that investment is higher in early life (0–2 years) and declines from 5–6 years.

Like MGNREGA the primary objective of the Expanded Public Work Program (EPWP) is to create temporary employment, especially in rural communities. The program was first set up in 2003 and further prioritized in NDP-2030. The NDP targeted to create 2 million temporary jobs under EPWP by 2030 (World Bank, 2018a, 2018b). The program has been applauded as one of the world's largest innovative interventions to protect people against hardship (Lotz-Sisitka, 2019). EPWP is a multidimensional program, it includes poverty alleviation, employment generation, skill development, infrastructure, and MSME development.

Maphanga and Mazenda (2019) study held that a stipend through EPWP is a useful instrument to reduce poverty, and makes families food secure. It also helps in improving skill set of participants. EPWP's expansion also contributed to early childhood development and home (Narayan, 2022). Antonopoulos and Kim's (2008) study shows the multiplying effect of EPWP program and cost-effectiveness.

Several literatures indicates that desired outcomes are yet to be realized. It is very difficult to find a permanent job after exiting temporary job under EPWP (Glen et al., 2019). Maphanga and Mazenda (2019) concluded that the lack in program implementation and coordination are major challenges. Mazenda et al., (2022) in a case study of Cape Town and Mangaung Metropolitan Municipality, highlighted the institutional challenges of EPWP program, such as weak reporting, lack of trained human resource, institutional capacity, inadequate funding, unstructured EPWP and political interference. Dladla (2023) in his doctoral thesis concluded that regulations of EPWP are flawed on the ground, such as paying the wage below the mandate of the program, external interference in the recruitment process, and weak planning and coordination from implementing agencies. Further, despite massive unemployment in South Africa, EPWP struggles to meet the mandated youth target. In an inauguration ceremony of the fifth phase of EPWP, Mr. Sihle Zikala, Minister of Public Work and Infrastructure, Republic of South Africa, highlighted the following challenges of EPWP including lack of coordination, ineffectiveness, lack of accountability, poor skill training, and absence of proper exit plan (OIS, 2023). Dreze (2022) notes similar challenges in MGNREGA, i.e. payment delay, wage stagnation in real terms, and corruption.

Agriculture: Source of Life and Livelihood

Agriculture is indispensable for sustainable development. A strong agriculture sector is key to achieving SDGs by improving living standards for all, especially poor and marginalized living in rural and backward areas (ADB, 2022). Agriculture is the source of livelihood for people living in rural areas (Niragira et al., 2019). According to World Bank Statistics (2024), nearly 32% of people in South Africa live in rural areas, while the corresponding figure for India is 64%, double that of South Africa.

Agriculture plays a critical role in economic growth and development. The agriculture sector has the potential to free people from poverty and hunger. It is more than a source of food, it creates opportunities, provides raw material to the industries, and contributes to the foreign reserves. In comparison to other sectors, agricultural growth has been 2–4 times more effective in increasing the poor's income (Townsend, 2015). Thus, investment in agriculture is vital for human well-being, especially for the mainstreaming of rural communities. So, it is enshrined in the political economy of India and South Africa.

The focus on agriculture in India has been evident since the Nehruvian era. Almost 20% of the public sector plan expenditure was assigned to the agriculture sector during the first and second five-year plans. Anyone can understand the importance of agriculture with the surge in credit flow to Rs. 2000 crore in 1975–76 from Rs. 70 crore in 1950–51 (Balakrishnan, 2007). During the Indo-Sino war emergency, at the annual meeting of the National Development Council of 1963, Nehru stated, "*Emergency merely does not mean soldiers, or getting aircraft. It means production, production for defense especially. All other types of production, more particularly on the agricultural front is equally necessary…we have done may things which are creditable to us. But overall picture is not of the fast progress, especially in agricultural domain. This is rather distressing because agriculture is the basis of all our development work. If we fail in agriculture, it does not matter what we achieve-how many plants we put up our economic development will not be complete. Agriculture is more important than industry because for the simple reason that industry depends on agriculture. Industry, which is, no doubt, very important will not progress unless agriculture is sound and stable and progressive*" (Publication Division, 1968).

Nehru's vision realized through the 'Green Revolution' transformed India into a food-secure country. India is among the leading producers of fruits and vegetables, the second world's largest in fruits and vegetables, contributing 9.3% to the total global production. At the same time, India is also among the few countries with higher post-harvest losses. Despite the impressive food production, the majority of farmers in India are unable to store their produce for a longer time. The monetary value of post-harvest loss in India is nearly Rs 926.51 billion, accounting for 40% of the total farm products of India (Tiwari et al., 2021). Subsequently, the monetary value of post-harvest loss pushes farmers into destitution.

The post-harvest infrastructure includes an integrated pack house, cold storage, reefer transport, and ripening houses. A high-level committee (Dalwi Committee) on doubling farmers' income in India revealed that India is grilling under a shortfall of agriculture infrastructure (Ministry of Agriculture, 2018). As shown in Table 5.1, 99.6% shortfall in the integrated packhouse, 85% of reefer transport, 91% repining stores, and 10% of cold storage.

Commercialization of agriculture is an ongoing process in India. The government has launched many schemes to assist and protect the farmers from externalities through various schemes such as *Kisan Samman Nidhi Yojana, Kisan Credit Cards, Sahakar Mitra Scheme, Pradhan Mantri Kisan Maan-Dhan Yojana*, e-*NAM* and many more.

The development of the agriculture sector is central to South Africa's policy-making. Liberalization of the agriculture sector was a prime objective for the policymakers in the post-apartheid period of South Africa. President Mandela believed that a strong and competitive agriculture sector in South Africa was essential to reap the benefits of globalization. In one of his famous speeches on the occasion of the KwaZulu-Natal Agriculture Union at Pietermaritzburg (RSA, 1997), he stated: "*As we promote a healthy, competitive agriculture sector we have left have left behind the days of protection by price guarantees, credit and inputs subsidies. Successful farmers, like their counterparts in industry, will be those who seize the opportunities of a competitive, global economy*".

The recent policy document of South Africa's ruling political party, African National Congress (ANC) outlined the importance of agriculture in unfitting people

Table 5.1 Availability of agriculture Infrastructure in India

Kind of agriculture Infrastructure	Required infrastructure	Existing infrastructure	Gap	Shortfall
Integrated packhouse	70,080	249	69,831	99.6
Reefer transport	61,826	<10,000	52,826	85
Cold storage	34,164,411	31,823,700	3,276,962	10
Cold storage	936,251			
Repining stores	9131	812	8319	91

Source Ministry of Agriculture, Government of India (2018)

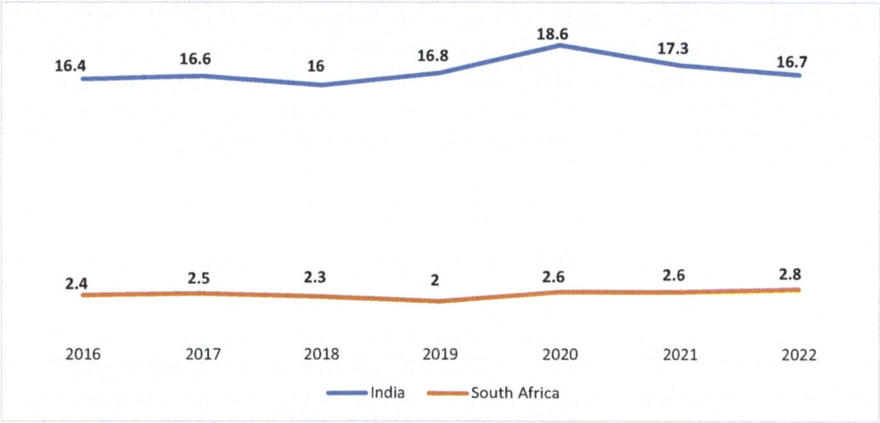

Fig. 5.1 Share of agriculture in total value addition in India and South Africa, 2016–2022. *Source* World Bank (2024a, 2024b, 2024c); Ministry of Agriculture, India (2023); Department of Agriculture, Land Reform and Rural Development, South Africa (2022)

from poverty (ANC, 2022). In consonance with the benefits of agriculture on overall development, the policy documented prioritized investment in the agriculture sector. Before ANC's recent policy document, the expansion of agriculture output and agri-business has been envisaged in South Africa's National Development Plan 2012.

Although the agriculture sector in South Africa is not a big contributor to the economy as in India, as shown in Fig. 5.1, it is an important contributor to the country's exports. The share of the agriculture sector in total exports has reached 11% in 2022 from an average of 9% in the last decades (Sihlobo, 2023), and almost 70% of agricultural produce is used as intermediate product (Dube et al., 2018). The low share of agriculture to GDP of South Africa highlights the need for more systematic policy research in context of structural change in the agriculture sector.

However, food wastage in South Africa is a major challenge for farmers and Policymakers. According to South Africa's Council of Scientific and Industrial Research, every year nearly 10.3 million tons of edible food are wasted in South Africa (CSIR, 2021). The share of lost food grains is 34% of local production and 45% of available food distribution in the country. The estimated cost of food loss is nearly ZAR 10 million.

Studies show that weak agricultural infrastructure is a major challenge in India (Hussain & Guha, 2023) and South Africa (Selepe et al., 2014; Samuel et al., 2019), and the situation has been worsened by climate change. World Bank (2013) revealed that the Indian summer monsoon would be high with an increase in 2° world's temperature. The report also found that South Africa has been facing human-induced warming since 1960.

The future of agriculture lies in the best adaptation. Investment in quality education, technology & innovation, skills, health, and sound infrastructure could lead to the path of sustainable agriculture. Asian Green Revolution shows that a 1% increase in agricultural output reduces poverty by 1.9% (Schroeder et al., 2021). India needs to share its vast experience of the green revolution as well as new technologies with South Africa through bilateral and multilateral forms like India-Africa Forum Summit, India Brazil, South Africa (IBSA), Brazil, Russia, India, China, and South Africa (BRICS), G-20, etc. to raise the productivity and income of farmers.

Health and Wellbeing: Key to Human Capabilities

Health is key to human life and seminal to human capabilities. The impact of good health can be observed on income, education, and women's empowerment. Health retardation blocks the prospect of human well-being. The causality between health and other dimensions of human development is explained by Michael Grossmann who figured that advancement in human productivity can be enhanced by better consumption of healthcare consumption (Grossman, 1972). World Bank (1993) equated public health with public good and warned against the negative consequences of bad health on individuals and households. Consequently, good health has the potential to accelerate economic growth.

WHO (2001) propagated the idea of investment in health. Sen (2002) identified health as an issue of social justice. Health finds its place in MDGs and is flagged as a priority in the SDGs. Inference from Chaps. 1, 2 and 3 shows the weak performance of India and South Africa behind the targets of MDGs, although some progress has been remarkable. Low investment in health infrastructure translated into inequitable access to health services, especially for the poor and marginalized. There is an urgent need in India and South Africa to strengthen their health sector and make it more equitable, otherwise as an African proverb says "the disease of a poor man takes him to his grave."

Chapter 4 inferred that, despite the low economic performance South Africa performed better than India in health indicators because most the individual expenditure is covered by the government. An individual in South Africa spent only just 5% from his pocket on health, while in India the share is 10 times higher, and the share of expenditure in India is nearly 50%.

Education: Gateway to Human Dignity

Since the beginning of civilization, education played a crucial role. Its importance is recognized by many philosophers. Laozi opined "*Give a man a fish he will eat for a day. Teach a man to fish he will eat for a life time*". Plato expounded "*If a man neglects education, he walks lame to the end of his life*", Mahatma Gandhi said, "*What is really needed to make democracy function is not knowledge of facts, but right to education*". Nobel laureate T. W. Schultz (1961) argued that spending on education translates into economic growth and Garry Becker (1992) framed the human capital theory. Nelson Mandela expressed that "*education is the weapon to fight against poverty*".

The vision and argument of aforesaid philosophers are reflected in the education policies of India and South Africa. *Right to Education Act, Mid-Day Meal* in India and *No-Fee Schools* in South Africa bring more children to schools and both countries have achieved significant progress in primary education. The Indian government's emphasis on quality education is visualized in the New Education Policy 2020 while South Africa has implemented a unique fee system to reduce the financial burden in higher education. The subsidized fee is only for the poor.

Investment in education yields fruitful results in the future. The rate of return from investment in education is about 9 to 10%, which can be understood as an addition of 10% in annual income with the addition of one more year of education (Patrinos, 2023). In developing countries, the quality of education depends on public expenditure on education. To achieve the target of quality education, UNESCO calls for efficient public spending on education (4 to 6% of GDP) (UNESCO, 2015). New Education Policy 2020 also calls for an increase in the education budget in India to 6%, still, public spending in India is just 4.5%, which is not enough to ensure inclusive quality education (see Fig. 5.2), whereas South Africa spent 6.6% of its GDP on education.

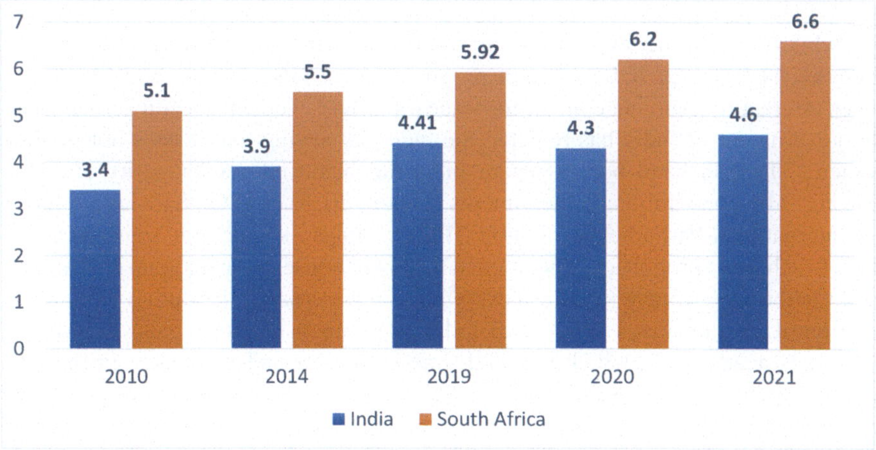

Fig. 5.2 Public spending on education (% of GDP), India and South Africa. *Source* World Bank (2024a, 2024b, 2024c), UNESCO (2024)

Gender Equality

The full participation of all citizens is indispensable to human development (UNDP, 2015a, 2015b). For far too long, chauvinisms and prejudices intrinsic to every culture and tradition across the world imposed deeply rooted structural systems that stripped away the rights and dignity of women as equal citizens (Singh, 2022a, 2022b). Biasness siege women's potential to fully participate in the economy (World Bank, 2023). Economies without the full participation of half the population cannot achieve their full potential. Equal contribution of men and women alike makes an economy more vibrant, robust, and resilient. Leaders and policymakers across the world have recognized the role of women in nation-building, policies have been enacted at several international forums and significant steps and measures have been implemented to achieve gender equality.

Global Gender Gap Report 2023 demonstrates that India ascended to eighth position and stood at 127th rank out of 146 countries from 135 in 2022 in terms of gender parity, whereas over the same period, South Africa remained at its earlier 20th position (WEF, 2023). According to statistics, South Africa performed better in each indicator of the index namely, access to economic opportunities, level of education, health and survival, and political participation at the national level.

Although India has achieved parity in educational attainment, many women are deprived of equal access to economic opportunity. Educated women also need decent jobs while most of the women end up doing excessive unpaid work at home. As discussed in Chap. 2, India's recent progress in sex ratio at birth and MMR has driven up equality. Another concerning issue for India is women's health and survival. The government of India has rejected the index by saying that it does not include women's share in local bodies where India has achieved gender parity after the 73rd and 74th constitutional amendments (Pandit, 2023). The Global Gender Index and International Parliamentary Union (IPU) readings indicate that South Africa is likely to achieve a proportion of women in the lower house of the national parliament, thanks to the 33% women's quota at every level of governance in South Africa; while in current Lok Sabha (lower house of Indian Parliament) the occupancy of women member is nearly 15%, which is an all-time high since independence. In the IPU ranking, South Africa is positioned at 14th, whereas India fell at 142nd out of 190 countries (IPU, 2024). Like South Africa, the Government of India has recognized the importance of women in decision-making and passed the long pending women's reservation bill that ensures 33% seats for women in the first and second tier of Indian democracy. However, its implementation is set to take another ten years.

Inter-Pandemic Period

The path of Sustainable Development Goals was drastically influenced by the COVID-19 pandemic. The impact of the pandemic was particularly harsh on the poor. The COVID-19 crisis has derailed the infrastructure of basic necessities and developing worlds have realized the immediate need to invest in health and other infrastructure. The pandemic propelled public debt levels and debt-service burdens, containing fiscal constrain. IMF (2023a, 2023b) projected the increasing debt ratio in developing countries and the debt ratio of India and South Africa will be increased to 81.9% and 73.7% respectively as shown in Fig. 5.3. The trends indicate that the public debt in South Africa will be doubled by 2028. The Governments of India and South Africa have responded with stimulus packages equivalent to 10% of their GDP to revitalize the economy and well-being of their people.

When I asked Jean Dreze what he considered to be the most fatal impact of the pandemic with regard to Human Development he pinpointed the devastation of the education system in developing countries. The marginalized and poor were hammered. Lack of smartphones and recharge money excluded them from the online education system. Students are promoted to upper grades without attaining the necessary basic knowledge. Children who have forgotten the Hindi alphabet because of the lockdown are now saddled with thick English textbooks on the reopening of schools.

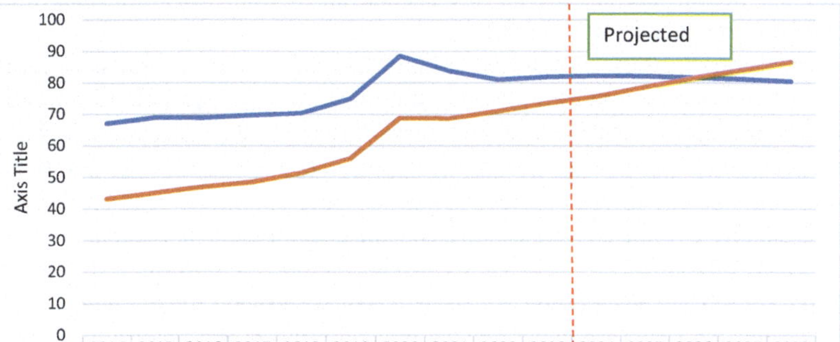

	2014	2015	2016	2017	2018	2019	2020	2021	2022	2023	2024	2025	2026	2027	2028
India	67.1	69	68.9	69.7	70.4	75	88.5	83.8	81	81.9	82.3	82.2	81.7	81.2	80.5
South Africa	43.3	45.2	47.1	48.6	51.5	56.1	68.9	68.8	71.1	73.7	75.8	78.8	81.6	84.2	86.7

Fig. 5.3 Public Debt (percentage of GDP), India and South Africa 2014–28. *Source* International Monetary Fund (2023)

Way Forward

Drawing an inference from previous chapters, I find that uneven development challenges need asymmetric strategies, contrary to the axiom that SDGs are indivisible. The idea of human development is enshrined in the long-term development agenda of *'Vikshit Bharat-2047*[1]' in India and *'Agenda 2063'* in South Africa. Saxena (2021) argues that federal governments face the unique challenge of maintaining a balance between the uneven states/provinces while ensuring health, education, food, shelter, and decent jobs through targeted pro-poor policies.

The heterogeneous structure of India and South Africa draws special attention to the social goals of SDGs: SDG1 (no poverty), SDG2 (zero hunger), SDG3 (health and wellbeing), SDG4 (quality education), SDG5 (gender equality), SDG 10 (inequality); environment goals: SDG6 (water and sanitation), SDG7 (affordable energy); and economic goals SDG 8 (decent jobs). The progress in social indicators is anticipated from better economic and environmental indicators.

In light of Chaps. 2 and 3, we can say that India and South Africa are countries with a large proportion of poor population over the small group of rich elites. The low saving or no saving limits the spending capacity of the former while the latter enjoy a high rate of saving, contracting their consumption. The success of SDGs in India and South Africa lies in the empowerment of the poor and marginalized. The respective countries need to unleash the full potential of human power by ensuring access to health services and quality education. By doing this India and South Africa transform its quantity into quality.

Undoubtedly, the COVID-19 pandemic has reversed the progress of human development, pushed people into poverty, and raised unemployment. The unprecedented challenges and after-effects of lockdown have created uncertainties. The solution to uncertainties lies in the Keynesian approach. Contrary to the Smithian idea of 'invisible hands', Keynes believed that, to some extent government intervention is necessary to regulate the market force because supply and demand in the market were not always in equilibrium. Keynes's model includes a multiplier, that is, the rate of output is multiple of spending. The rate of return is always greater than spending if the fiscal multiplier is more than one (Jahan, Mahmud, & Papageorgiou, 2014). Keynes argued that the aggregate consumer demand can be boosted by purchasing power derived from better job opportunities, production, and consumption (see illustration 5.1) (Keynes, 2018).

The return to spending on infrastructure will significantly increase when health and education receive similar levels of investment. The complementary investment in health and education will enhance the rate of return, by improving employment opportunities and well-being. I have attempted to forge the Keynesian approach with Sen's capability approach. By doing this, I propose a new model to shift the economy on the path of sustainable development (see illustration 5.2). The model finds its genesis in Amartya Sens's concept of capability approach. To achieve development

---

[1] Vikshit Bharat-2047 is the vision to transform India in to a developed nation by 2047. The vison entails inclusive development through good governance.

Illustration 5.1 Growth model based Keynesian approach. *Source* Illustration Arranged by Ravi Tomar

objectives leaders and policymakers must ensure 3R's, reach (reach of the target to be accomplished), range (tools to used), and region (significance to pursue). The poor and marginalized face challenges in expanding their choices, they are not free to realize their full potential. It is the duty of the state to ensure equal and unbiased opportunities for all, they are free to do or to be.

I believe that the empowerment of an individual is dependent on human capital. The level of education and well-being depends on decent income generated from employment opportunities.

On the whole, the book has established the causalities between indicators of human development and SDGs and the outcome can be used in the analysis of Government spending on the level of human development in the framework of Sustainable Development Goals, as well as policy intervention in development strategies. It has also been identified that every country is unique so are their challenges. Growth should not be limited to a best practice model, it should be dynamic, with the capability of change following local needs and challenges.

To some extent, this study addresses the puzzle; despite the high growth rate, why is India's performance lower than South Africa's in terms of social indicators? It does so by drawing attention to welfare approaches. Investment in drivers of human capital in India is lower than in South Africa. Welfare programs address the needs of the people, like subsidized food to nearly 60% Indian population, their options are

Illustration 5.2 Proposed model to achieve human development. *Source* Illustration Arranged by Ravi Tomar

limited, while the financial grant in South Africa gives freedom to choose the better option. Also, the public spending on health and education in South Africa is greater than in India. Policymakers and leaders of India and South Africa must understand that humans are the true wealth of an economy, the success of any development goals is impossible without the freedom to choose the options essential to human prosperity.

The finding suggests that with improvements in health, education, and income, a rise in purchasing power can be anticipated on the demand side. On the supply side, public spending on education and health is indispensable for the enhancement of human capabilities. Both governments face the unique challenge of maintaining a balance between the uneven states/provinces while ensuring health, education, food, shelter, and decent jobs through targeted pro-poor policies. Hence, the programs and policies of India and South Africa must further focus on the need to strengthen the determinants of human development to attain the Sustainable Development Goals by 2030.

References

AbhijitBanerjee, EstherDuflo, & GarimaSharma. (2021). *Long-term effects of the targeting the ultra poor program.*. Avialable at: https://www.nber.org/system/files/working_papers/w28074/w28074.pdf: NATIONAL BUREAU OF ECONOMIC RESEARCH.

ADB. (2022). *ADB's work in agriculture and food security.*. https://www.adb.org/what-we-do/topics/agriculture/overview: Asian development Bank.

Ananian, S., & Dellaferrera, G. (2024). *Employment and wage disparities between rural and urban areas.* https://www.ilo.org/wcmsp5/groups/public/---dgreports/---inst/documents/publication/wcms_911918.pdf: International Labour Organization.

ANC. (2022). *Umrabulo: policy conference 2022 special edition.*. Available at: https://www.anc1912.org.za/wp-content/uploads/2022/09/6th-ANC-National-Policy-Conference-Reports-2022.pdf: African National Congress.

Antonopoulos, R., & Kim, K. (2008). *Research Project No. 34. The impact of public employment guarantee strategies on gender equality and pro-poor economic development.*. Levy Economics Institute of Bard College. https://www.levyinstitute.org/pubs/rpr_01_08_SA.pdf.

Banerjee, A. V., & Newman, A. F. (1993). Occupational choice and the process of development. *Journal of Political Economy, 101*(2), 274–298. Available at: https://www.jstor.org/stable/2138820.

Banerjee, A., Hanna, R., Olken, B. A., & Lisker, D. S. (2023). Social protection in the developing world. *Massachusetts Institute of Technology*, 1–78. Available at: https://economics.mit.edu/sites/default/files/2023-08/Social_Protection_paper_manuscript.pdf.

Balakrishnan, P. (2007). The Recovery of India: Economic Growth in the Nehru Era. *Economic & Political Weekly, Vol. 42, Issue No. 45–46*, 52–66.

Barrientos, A. (2013). *Social assistance in developing countries.* Cambridge University Press.

Basole, A., Abraham, R., Lahoti, R., Kesar, S., Jha, M., Nath, P., … Narayanan, R. (2021). *State of working India 2021: One year of Covid-19.*. Available at: https://azimpremjiuniversity.edu.in/publications/2024/report/state-of-working-india-2021-one-year-of-covid-19: Azim Premji University.

Becker, U. (2014). *The BRICs and emerging economies in comparative perspective: Political economy.* Routledge Taylor & Francis eBooks.

Berg, E., Bhattacharyya, S., Rajasekhar, D., & Manjula, R. (2018). Can public works increase equilibrium wages? Evidence from India's National Rural Employment Guarantee. *World Development, 103*, 239–254. Available at: https://doi.org/10.1016/j.worlddev.2017.10.027.

Bharadwaj, P., Lakdawala, L. K., & Li, N. (2019). Perverse consequences of well intentioned regulation: Evidence from India's child labor ban. *Journal of the European Economic Association, 18*(3), 1158–1195. Available at: https://doi.org/10.1093/jeea/jvz059.

Chetty, R., & Looney, A. (2006). Consumption smoothing and the welfare consequences of social insurance in developing economies. *Journal of Public Economics, 90*(12), 2351–2356. Available at: https://doi.org/10.1016/j.jpubeco.2006.07.002.

CSIR. (2021). *45% of available food supply in south Africa wasted, shows new CSIR study.*

Dadush, U., & Shaw, W. (2011). *Juggernaut: How emerging markets are reshaping globalization.* Carnegie Endowment for International Peace.

Deininger, K., & Liu, Y. (2019). Heterogeneous welfare impacts of national rural employment guarantee scheme: Evidence from Andhra Pradesh, India. *World Development, Elsevier, 117(C)*, 98–111.

Department of Agriculture, Land Reform and Rural Development. (2022). *Economic review of the South African agriculture.* Department of Agriculture, Land Reform and Rural Development, South Africa.

Dladla, L. G. (2023). *Public employment programmes and their contribution to service delivery and rural livelihoods in South Africa.* KWAZULU-NATAL. Avaialble at: https://researchspace.ukzn.ac.za/server/api/core/bitstreams/b1ff7f7c-0619-4715-84e6-e3c1843b246a/content: UNIVERSITY OF KWAZULU-NATAL.

Drèze, J. (2022). *Employment guarantee in action: Insights from India.* Available at: EDI%20Note%20Employment%20Guarantee%20in%20Action%20Insights%20from%20India.pdf: Economic Democracy Initiative.

Dube, S., Nair, R. D., Nkhonjera, M., & Tempia, N. (2018). *Structural transformation in agriculture and agro-processing value chains. CCRED Working Paper No. 8/2018.* Centre for Competition, Regulation and Economic Development (CCRED) Working Paper Series.

Etana, D., & Tolossa, D. (2017). Unemployment and food insecurity in urban Ethiopia. *African Development Review, 29*(1), 56–68. Available at: https://doi.org/10.1111/1467-8268.12238.

FAO. (2020). *Digital agriculture profile: South Africa.* Available at: https://www.fao.org/3/cb2506en/CB2506EN.pdf: Food and Agriculture Organizations.

Fetzer, T. (2020). Can workfare programs moderate conflict? Evidence from India. *Journal of the European Economic Association, 18*(6), 3337–3375. Available at: https://doi.org/10.1093/jeea/jvz062.

Gallie, D., Paugam, S., & Jacobs, S. (2010). Unemployment, poverty and social isolation: Is there a vicious circle of social exclusion? *European Societies, 5, 2003*(1), 1–32. Available at: https://doi.org/10.1080/1461669032000057668.

Gehrke, E. (2019). An employment guarantee as risk insurance? Assessing the effects of the NREGS on agricultural production decisions. *The World Bank Economic Review, 33*(2), 413–435. Available at: https://doi.org/10.1093/wber/lhw067.

Gentilini, U., Almenfi, M., Iyengar, H. T., Okamura, Y., Downes, J. A., Dale, P., … Vi. (2022). *Social Protection and Jobs Responses to COVID-19 : A Real-Time Review of Country Measures.* Available at: https://documents1.worldbank.org/curated/en/110221643895832724/pdf/Social-Protection-and-Jobs-Responses-to-COVID-19-A-Real-Time-Review-of-Country-Measures.pdf: World Bank.

Gigineishvili, N., Teodoru, I. R., Karapetyan, N., Ustyugova, Y., Houtte, J. v., Jonas, J., … Saliba, F. (2023). Strengthening social safety nets. In N. Gigineishvili, I. R. Teodoru, N. Karapetyan, Y. Ustyugova, J. V. Houtte, J. Jonas, … F. Saliba, *Paving the Way to More Resilient, Inclusive, and Greener Economies in the Caucasus and Central Asia* (pp. 47–63). International Monetary Fund. Available at: https://doi.org/10.5089/9798400239175.087.

Glen, N., Fourie, K., & Moche, N. (2019). Rural Tourism development: Exit opportunities for expanded public works participants. *Journal of Advanced Management Science, 7*(3), 84–93. Available at: https://www.joams.com/uploadfile/2019/0829/20190829051323328.pdf.

Grebmer, K. V., Bernstein, J., Wiemers, M., Reiner, L., Bachmeier, M., Hanano, A., … Fritschel, H. (2023). *Global hunger index-2023.* WHH.

Grossman, M. (1972). On the concept of health capital and the demand for health. *Journal of Political Economy*, 223–255.

Hanna, R., & Karlan, D. (2017). Designing social protection programs: using theory and experimentation to understand how to help combat poverty. In E. D. Abhijit Vinayak Banerjee, *Handbook of Economic Field Experiments. Volume 2* (pp. 515–553). Available at: https://doi.org/10.1016/bs.hefe.2016.07.002.

Hussain, M. A., & Guha, P. (2023). Role of farm infrastructure in agribusiness during a crisis: insights from rural Assam, India. *7*, 1035–1054. https://doi.org/10.1007/s41685-023-00304-8.

Imbert, C., & Papp, J. (2019). Short-term migration, rural public works, and urban labor markets: Evidence from India. *Journal of the European Economic Association, 18*(2), 927–963. https://doi.org/10.1093/jeea/jvz009

IMF. (2023a). *Fiscal monitor: Climate crossroads: Fiscal policies in a warming world.* Available at: https://www.imf.org/en/Publications/FM/Issues/2023/10/10/fiscal-monitor-october-2023: International Monetary Fund.

IMF. (2023b). *World economic outlook.* Available at: https://www.imf.org/external/datamapper/PPPSH@WEO/OEMDC/ADVEC/WEOWORLD: International Monetary Fund.

IPU. (2024). *Ranking as of 1st February 2024.*

Islam, M., & Sivasankaran, A. (2015). *How does child labor respond to changes in adult work opportunities? Evidence from NREGA*. Harvard. Available at: https://scholar.harvard.edu/files/mahnazislam/files/nrega_mahnaz_anitha_1.pdf: Harvard University.

Joshi, M. K., & Klein, J. (2018). Emerging Economies as Growth Drivers. In M. Joshi, & J. R. Klein, *Global Business* (pp. 56–62). Oxford Academic: Available at: https://doi.org/10.1093/oso/9780198827481.003.0008.

Leoni, S. (2023). A historical review of the role of education: from human capital to human capabilities. *Review of Political Economy*, Available athttps://doi.org/10.1080/09538259.2023.2245233

Li, T., & Sekhri, S. (2020). The spillovers of employment guarantee programs on child labor and education. *The World Bank Economic Review, 34*(1), 164–178. https://doi.org/10.1093/wber/lhy027

Liu, M. (2021). Politics and governance in emerging market countries. *Global Journal of Emerging Market Economies, 13*(2), 265–286. Available at: https://doi.org/10.1177/09749101211004378.

Lotz-Sisitka, H. (2019). Probing the potential of social ecosystemic skills approaches for green skills planning. In E. Rosenberg, P. Ramsarup, & H. Lotz-Sisitka (Eds.), *Green Skills Research in South Africa* (pp. 112–127). Taylor & Francis.

Maphanga, M., & Mazenda, A. (2019). The effectiveness of the expanded public works programme as a poverty alleviation strategy. *Administratio Publica, 27*(3), 7–26. Available at: https://repository.up.ac.za/bitstream/handle/2263/71497/Maphanga_Effectiveness_2019.pdf?sequence=1.

Mazenda, A., Matjane, K., Simawu, M., & Lubinga, S. (2022). Public works as a safety Net: Evidence and implementation in South African municipalities. *Urban Affairs and Public Policy, 23*, 33–49. Available at: https://doi.org/10.47785/urbana.1.2022.3.

MECS. (2023). *Modern energy cooking brief: South Africa.* Leicestershire. Available at: https://mecs.org.uk/wp-content/uploads/2023/06/South-Africa.pdf: Loughborough University.

Ministry of Agriculture. (2023). *Annula Report-: 2022–23.* Ministry of Agriculture. Governmnet of India.

Ministry of Agriculture. (2018). *Report of the committee on doubling farmers' income: Comprehensive policy recommendations".* Available in India: https://farmer.gov.in/imagedefault/DFI/DFI_Volume_14.pdf: Misitry of Agriculture, Government of India.

Narayan, S. (2022). Fifteen Years of India's NREGA: Employer of the last resort? *The Indian Journal of Labour Economics, 65*, 779–799. Available at: https://doi.org/10.1007/s41027-022-00396-4.

Nilsson, A. (2015). Who suffers from unemployment? The role of health and skills. *IZA Journal of Labor Policy, 4, Article number: 19*, Available at: https://doi.org/10.1186/s40173-015-0046-5.

Niragira, S., Ndimubandi, J., & Orshoven, J. V. (2019). Income, time and labor nexus household food security in Burundi. *Encyclopedia of Food Security and Sustainability, 1*, 534–539. Available at: https://doi.org/10.1016/B978-0-08-100596-5.22375-5.

Nuruzzaman, M. (2020). Why BRICS is no threat to the post-war liberal world order. *International Studies, 57*(1), 51–66. Available at: https://doi.org/10.1177/0020881719884449.

OIS. (2023). Minister Sihle Zikalala: EPWP Phase 5 Indaba. *Official Information and Services*, pp. Available at: https://www.gov.za/news/speeches/minister-sihle-zikalala-epwp-phase-5-indaba-10-oct-2023.

Pandit, A. (2023). WEF recognises need to record women's participation in local govt bodies in gender gap report. *Times of India*, pp. https://timesofindia.indiatimes.com/india/wef-recognises-need-to-record-womens-participation-in-local-govt-bodies-in-gender-gap-report/articleshow/97259389.cms.

Patrinos, H. A. (2023). 50 years after landmark study, returns to education remain strong. *Published on Education for Global Development*, pp. https://blogs.worldbank.org/education/50-years-after-landmark-study-returns-education-remain-strong#:~:text=But%20education%20is%20more%20than,10%25%20increase%20in%20earnings%20annually.

PIB. (2023). *Health an integral component of social welfare, leitmotif for the government.*. Press Information Bureau, Govermnet of India.

Pohlan, L. (2019). Unemployment and social exclusion. *Journal of Economic Behavior & Organization, 164*, 273–299. Available at: https://doi.org/10.1016/j.jebo.2019.06.006.

Psacharopoulos, G., & Patrinos, H. A. (2023). *Returns to investment in education a decennial review of the global literature*. World Bank. Available at: https://openknowledge.worldbank.org/server/api/core/bitstreams/cb09c589-14c6-5675-8026-39333ff71532/content.

Publication Division. (1968). *Jawahar Lal Nehru's Speaches: March 1963 to May 1964.*. Available at: https://ignca.gov.in/Asi_data/48033.pdf: Public Divison, Ministry of Information and Broadcasting. Government of India.

PWC. (2017). *The Long View How will the global economic order change by 2050?* .. PWC. Available at: https://www.pwc.com/gx/en/world-2050/assets/pwc-world-in-2050-summary-report-feb-2017.pdf.

R.L., J., C.P., S., & T.J., S. (1995). The impact of unemployment on health: A review of the evidence. *CMAJ. Canadian Medical Association Journal, 153*(5), 529–540. Available at: https://www.scopus.com/record/display.uri?eid=2-s2.0-0029134582&origin=inward&txGid=d7d72e6f911d97bc2f1b0db4d3a5f0c0.

Rodrik, D. (2024). Addressing challenges of a new era: Against rule-of-thumb economics. *Finance & Development*, pp. https://www.imf.org/en/Publications/fandd/issues/2024/03/Point-of-view-addressing-challenges-of-a-new-era-Dani-Rodrick?utm_medium=email&utm_source=govdelivery.

RSA. (1997). *Condolence book*. Pretoria. Available at: http://www.mandela.gov.za/mandela_speeches/1997/970424_kznagric.htm: Republic of South Africa.

S., R. (2021). MGNREGS reduced poverty, empowered women. *The Hindu*, pp. https://www.thehindu.com/data/MGNREGA-reduced-poverty-empowered-women-NCAER/article60321034.ece#:~:text=The%20programme%20reduced%20poverty%20by,people%20from%20falling%20into%20poverty.

Samuel, O. O., Sylvia, T. S., & Casadevall, S. R. (2019). Establishing the nexus between climate change adaptation strategy and smallholder farmers' food security status in South Africa: A bi-casual effect using instrumental variable approach. *Cogent Social Sciences, 5*(1), Available at: https://doi.org/10.1080/23311886.2019.1656402.

Saxena, R. (2021). The working of cooperative and collaborative federalism in India: Understanding intergovernmental relations. *Indian Journal of Public Administration*, 153–164. Available at: https://doi.org/10.1177/00195561211026621.

Schroeder, K., Lampietti, J., & Elabed, G. (2021). *What's cooking: Digital transformation of the agrifood system*. Available athttps://doi.org/10.1596/978-1-4648-1657-4: WorldBank

Selepe, M., Sabela, T., & Masuku, M. (2014). The effect of infrastructural challenges on food security in Ntambanana, KwaZulu-Natal, South Africa. *African Journal of Hospitality, Tourism and Leisure, 3*(1), Available at: https//:www.ajhtl.com.

Sen, A. (2002). Why health equity? *Health Economics*, 659–666.

Shah, M., & Steinberg, B. M. (2017). Drought of opportunities: Contemporaneousand long-term impacts of rainfall shocks on human capital. *Journal of Political Economy, 125*(2), 527–561.

Sihlobo, W. (2023). South Africa is exporting more food. But it needs to find new growth frontiers. *The Conversation*, pp. https://theconversation.com/south-africa-is-exporting-more-food-but-it-needs-to-find-new-growth-frontiers-200747.

Singh, U. K. (2022). India's poorest states need special healthcare focus—one size Won't fit all. *The Qunit*, pp. https://www.thequint.com/opinion/indias-poorest-states-need-special-healthcare-focus-one-size-doesnt-fit-all#read-more.

Singh, U. K. (2022). Voluntary party quota: An alternative route to achieve women's representation. *The Quint*, pp. https://www.thequint.com/gender/women-reservation-in-politics-alternate-route-quota-in-parties.

Srivastava, S. (2012). The emerging economies and changing prospects of the multicurrency global order: avenues and challenges in times ahead. *Procedia—Social and Behavioral Sciences, 37*, 46–56. Available at: https://doi.org/10.1016/j.sbspro.2012.03.274.

STATS SA. (2021). *Assessing food inadequacy and hunger in South Africa in 2021 using the General Household Survey (GHS).* Available at: https://www.statssa.gov.za/publications/03-00-20/03-00-202021.pdf: Department of Statistics. Republic ofSouth Africa.

STATS SA. (2023). *Macro-social report-2022.* Availabe at: https://mistra.org.za/wp-content/uploads/2023/05/Macro-Social-Report-2023-ONLINE-VERSION-24-MAY-compressed.pdf: Stats SA.

Tillin, L., & Duckett, J. (2017). The politics of social policy: welfare expansion in Brazil, China, India and South Africa in comparative perspective. *Commonwealth & Comparative Politics, 55*(3), 253–277. Available at: https://doi.org/10.1080/14662043.2017.1327925.

Tiwari, A., Afroz, S. B., & Kumar, V. (2021). Market vulnerabilities and potential of horticulture crops in India: With special reference to top crops. *Indian Journal Agricultural Markting, 35*(3).

Townsend, R. F. (2015). *Ending poverty and hunger by 2030: An agenda for the global food system.* Avaialble at: https://documents1.worldbank.org/curated/en/700061468334490682/pdf/95768-REVISED-WP-PUBLIC-Box391467B-Ending-Poverty-and-Hunger-by-2030-FINAL.pdf: World Bank.

UN. (2022). Nutrition and food security. *Nutrition and food security*, pp. Available at: https://india.un.org/en/171969-nutrition-and-food-security#:~:text=The%20share%20of%20agriculture%20and,in%20the%20last%2030%20years.

UN. (2023). *Global multidimensional poverty index 2023.* Available at: https://hdr.undp.org/content/2023-global-multidimensional-poverty-index-mpi#/indicies/MPI: United Nations.

UNDP. (2015a). *Gender equality in human development—measurement revisited.* United Nations Development Programme

UNDP. (2015b). *Human development report 2015.* United Nations Development Program.

UNESCO. (2015). *Incheon declaration and framework for action.* UNESCO.

UNESCO. (2024). *Sustainable development goals : Government expenditure on education as a percentage of GDP.* http://data.uis.unesco.org/index.aspx?queryid=3865: UNESCO.

UNICEF. (2023). *Maternal mortality declined by 34 per cent between 2000 and 2020.* Avaialble at: https://data.unicef.org/topic/maternal-health/maternal-mortality/: UNICEF.

Vaalavuo, M., & Sirniö, O. (2021). Obs against poverty: a fixed-effects analysis on the link between gaining employment and exiting poverty in Europe. *European Societies, 24*(4), 431–462. Availabke at: https://doi.org/10.1080/14616696.2022.2088821.

WEF. (2023). *Global gender gap report 2023.* World Economic Forum.

Wheebox. (2024). *India skills report 2024.* Gurugram. Avialable at: https://wheebox.com/assets/pdf/ISR_Report_2024.pdf: Wheebox.

WHO. (2001). *Macroeconomics and health: Investing in health for economic development. Report of the Commission on Macroeconomics and Health.* World Health Organization.

Widrick, S. (2007). *Emerging global markets: A Five-country comparative study Emerging global markets: A Five-country comparative study.* https://repository.rit.edu/books/9?utm_source=repository.rit.edu%2Fbooks%2F9&utm_medium=PDF&utm_campaign=PDFCoverPages.

Wolla, S. A., & Vandenbroucke, G. (2023). *Is College Still Worth the High Price? Weighing Costs and Benefits of Investing in Human Capital.* Page One Economics®. Available at: https://files.stlouisfed.org/files/htdocs/publications/page1-econ/2023/09/01/is-college-still-worth-the-high-price-weighing-costs-and-benefits-of-investing-in-human-capital_SE.pdf.

World Bank. (2018a). *The state of social safety nets 2018.* Available athttps://doi.org/10.1596/978-1-4648-1254-5:WorldBank

World Bank. (2018b). *An incomplete transition overcoming the legacy of exclusion in South Africa.* Available at: https://documents1.worldbank.org/curated/en/815401525706928690/pdf/WBG-South-Africa-Systematic-Country-Diagnostic-FINAL-for-board-SECPO-Edit-05032018.pdf: World Bank.

World Bank. (1993). *World development report 1993: investing in health.* Oxford University Press. World Bank.

World Bank. (2012). *Resilience, equity, and opportunity : the World Bank's social protection and labor strategy 2012–2022.* Available at: http://documents.worldbank.org/curated/en/443791 468157506768/Resilience-equity-and-opportunity-the-World-Banks-social-protection-and-labor-strategy-2012-2022: World Bank.

World Bank. (2013). *Turn down the heat: Climate extremes, regional impacts, and the case for resilience.* World Bank.

World Bank. (2019). Safety Nets: ocial safety net programs protect families from the impact of economic shocks, natural disasters, and other crises. *Understanding Poverty*, p. Available at: https://www.worldbank.org/en/topic/safetynets.

World Bank. (2022). *Correcting curse: Poverty and shared properity-2022* . World Bank.

World Bank. (2023). *Women, business and the law.*. World Bank.https://doi.org/10.1596/978-1-4648-1944-5

World Bank. (2024). *Agriculture, forestry, and fishing, value added (% of GDP)—South Africa, India.*. Available at: https://data.worldbank.org/indicator/NV.AGR.TOTL.ZS?locations=ZA-IN: World Bank.

World Bank. (2024). *Government expenditure on education, total (% of GDP).*. https://genderdata.worldbank.org/indicators/se-xpd-totl-gd-zs/?year=2022: World Bank.

World Bank. (2024). *Rural population (% of total population)—India, South Africa.*. Availabe at: https://data.worldbank.org/indicator/SP.RUR.TOTL.ZS?locations=IN-ZA: World Bank.

Wu, S., & Pan, Q. (2021). Economic growth in emerging market countries. *Global Journal of Emerging Market Economies, 13*(2), 192–215. Available at: https://doi.org/10.1177/097491012 11004405.

GPSR Compliance

The European Union's (EU) General Product Safety Regulation (GPSR) is a set of rules that requires consumer products to be safe and our obligations to ensure this.

If you have any concerns about our products, you can contact us on ProductSafety@springernature.com

In case Publisher is established outside the EU, the EU authorized representative is:

Springer Nature Customer Service Center GmbH
Europaplatz 3
69115 Heidelberg, Germany

The manufacturer's authorised representative in the EU is Springer Nature Customer Service Centre GmbH, Europaplatz 3, 69115 Heidelberg, Germany. If you have any concerns regarding our products, please contact ProductSafety@springernature.com

Printed and bound by CPI Group (UK) Ltd, Croydon, CR0 4YY
01/10/2026
02225923-0004